Science and Religion,
400 B.C. to A.D. 1550

Science and Religion, 400 B.C. to A.D. 1550

From Aristotle to Copernicus

EDWARD GRANT

Greenwood Guides to Science and Religion
Richard Olson, Series Editor

Greenwood Press
Westport, Connecticut • London

Library of Congress Cataloging-in-Publication Data

Grant, Edward, 1926–
   Science and religion, 400 B.C. to A.D. 1550 : from Aristotle to Copernicus /
Edward Grant.
     p. cm.—(Greenwood guides to science and religion)
   Includes bibliographical references and index.
   ISBN 0–313–32858–7 (alk. paper)
   1. Religion and science—History. I. Title. II. Series.
 BL240.3.G74 2004
  201′.65′09—dc22   2004017429

British Library Cataloguing in Publication Data is available.

Library of Congress Catalog Card Number: 2004017429
ISBN: 0–313–32858–7

First published in 2004

Greenwood Press, 88 Post Road West, Westport, CT 06881
An imprint of Greenwood Publishing Group, Inc.
www.greenwood.com

Printed in the United States of America

The paper used in this book complies with the
Permanent Paper Standard issued by the National
Information Standards Organization (Z39.48–1984).

10 9 8 7 6 5 4 3 2 1

Copyright Acknowledgment

To my granddaughters
Hannah Xin-Li Grant
and
Sarah Jia-Li Grant
With all my love
Grandpa

Contents

Illustrations

Series Foreword

For nearly 2,500 years, some conservative members of society have expressed concern about the activities of those who sought to find a naturalistic explanation for natural phenomena. In 429 B.C.E., for example, the comic playwright Aristophanes parodied Socrates as someone who studied the phenomena of the atmosphere, turning the awe-inspiring thunder that had seemed to express the wrath of Zeus into nothing but the farting of the clouds. Such actions, Aristophanes argued, were blasphemous and would undermine all tradition, law, and custom. Among early Christian spokespersons, there were some, such as Tertullian, who also criticized those who sought to understand the natural world on the grounds that they "persist in applying their studies to a vain purpose, since they indulge their curiosity on natural objects, which they ought rather [direct] to their Creator and Governor."[1]

In the twentieth century, though a general distrust of science persisted among some conservative groups, the most intense opposition was reserved for the theory of evolution by natural selection. Typical of extreme anti-evolution comments is the following opinion offered by Judge Braswell Dean of the Georgia Court of Appeals: "This monkey mythology of Darwin is the cause of permissiveness, promiscuity, pills, prophylactics, perversions, pregnancies, abortions, pornography, pollution, poisoning, and proliferation of crimes of all types."[2]

It can hardly be surprising that those committed to the study of natural phenomena responded to their denigrators in kind, accusing them of willful ignorance and of repressive behavior. Thus, when Galileo Galilei was warned against holding and teaching the Copernican system of astronomy as true, he wielded his brilliantly ironic pen and threw down a gauntlet to religious authorities in an introductory letter, "To the Discerning Reader," at the beginning of his great *Dialogue Concerning the Two Chief World Systems*:

Several years Ago there was published in Rome a salutory edict which, in order to obviate the dangerous tendencies of our age, imposed a seasonable silence upon the Pythagorean [and Copernican] opinion that the earth moves. There were those who impudently asserted that this decree had its origin, not in judicious inquiry, but in passion none too well informed. Complaints were to be heard that advisors who were totally unskilled at astronomical observations ought not to clip the wings of reflective intellects by means of rash prohibitions.

Upon hearing such carping insolence, my zeal could not be contained.[3]

No contemporary discerning reader could have missed Galileo's anger and disdain for those he considered enemies of free scientific inquiry.

Even more bitter than Galileo was Thomas Henry Huxley, often known as "Darwin's bulldog." In 1860, after a famous confrontation with the Anglican bishop Samuel Wilberforce, Huxley bemoaned the persecution suffered by many natural philosophers, but then he reflected that the scientists were exacting their revenge:

Extinguished theologians lie about the cradle of every science as the strangled snakes beside that of Hercules; and history records that whenever science and orthodoxy have been fairly opposed, the latter has been forced to retire from the lists, bleeding and crushed, if not annihilated; scotched if not slain.[4]

The impression left, considering these colorful complaints from both sides, is that science and religion must continually be at war with one another. That view was reinforced by Andrew Dickson White's *A History of the Warfare of Science with Theology in Christendom*, which has seldom been out of print since it was published as a two-volume work in 1896. White's views have shaped lay understanding of science and religion interactions for more than a century, but recent and more careful scholarship has shown that confrontational stances do not represent the views of the overwhelming majority of either scientific investigators or religious figures throughout history.

One response among those who wish to deny that conflict constitutes the most frequent relationship between science and religion is to claim such conflict cannot exist because these pursuits address completely different human needs and therefore have nothing to do with one another. This was the position of Immanuel Kant who insisted that the world of natural phenomena, with its dependence on deterministic causality, is fundamentally disjoint from the noumenal world of human choice and morality, which constitutes the domain of religion. Much more recently, it was the position taken by Stephen Jay Gould in *Rocks of Ages: Science and Religion in the Fullness of Life*:

I . . . do not understand why the two enterprises should experience any conflict. Science tries to document the factual character of the natural world and to develop theories that coordinate and explain these facts. Religion, on the other hand, operates in the equally important, but utterly different realm of human purposes, meanings, and values.[5]

In order to capture the disjunction between science and religion, Gould enunciates a principle of "Non-overlapping magisterial," which he identifies as "a principle of respectful noninterference."[6]

In spite of the intense desire of those who wish to isolate science and religion from one another in order to protect the autonomy of one, the other, or both, there are many reasons to believe that this is ultimately an impossible task. One of the central issues addressed by many religions is the relationship between members of the human community and the natural world. This is a central question addressed in Genesis, for example. Any attempt to relate human and natural existence depends heavily on the understanding of nature that exists within a culture. So where nature is studied through scientific methods, scientific knowledge is unavoidably incorporated into religious thought. The need to interpret Genesis in terms of the dominant understandings of nature thus gave rise to a tradition of scientifically informed commentaries on the six days of creation, which constituted a major genre of Christian literature from the early days of Christianity through the Renaissance.

It is also widely understood that in relatively simple cultures—even those of early urban centers—there is a low level of cultural specialization, so economic, religious, and knowledge-producing specialties are highly integrated. In Bronze-Age Mesopotamia, for example, agricultural activities were governed both by knowledge of the physical conditions necessary for successful farming and by religious rituals associated with plowing, planting, irrigating, and harvesting. Thus, religious practices and natural knowledge interacted to establish the character and timing of farming activities.

Even in very complex industrial societies with high levels of specialization and division of labor, the various cultural specialties are never completely isolated from one another and they share many common values and assumptions. Given the linked nature of virtually all institutions in any culture, it is the case that when either religious or scientific institutions change substantially, those changes are likely to produce pressure for change in the other. It was probably true, for example, that the attempts of pre-Socratic investigators of nature, with their emphasis on uniformities in the natural world and apparent examples of events systematically directed toward particular ends, made it difficult to sustain beliefs in the old pantheon of human-like and fundamentally capricious Olympian gods. But it is equally true that the attempts to understand nature promoted a new notion of the divine—a notion that was both monotheistic and transcendent, rather than polytheistic and immanent—that focused on both justice and intellect rather than power and passion. Thus, early Greek natural philosophy undoubtedly played a role not simply in challenging but also in transforming Greek religious sensibilities.

Transforming pressures do not always run just from scientific to religious domains, however. During the Renaissance, there was a dramatic change of thought among Christian intellectuals from one that focused on the contemplation of God's works to one that focused on the responsibility of the Christian to care for his fellow humans. The active life of service to humankind, rather than the contemplative life of reflection on God's character and works, now became the Christian ideal for many. As a consequence of this new focus on the active life, Renaissance intellectuals turned away from the then dominant Aristotelian view of science, which saw the inability of theoretical sciences to change the world as a positive virtue. They replaced this understanding with a new view of natural knowledge, promoted in the writings of men such as Johann Andreae in Germany and Francis Bacon in England, which viewed natural knowledge as significant only because it gave humankind the ability to manipulate the world to improve the quality of life. Natural knowledge would henceforth be prized by many because it conferred power over the natural world.

Modern science thus took on a distinctly utilitarian shape, a response due at least in part to religious changes.

Neither the conflict model nor the claim of disjunction, then, accurately reflect the often intense and frequently supportive interactions between religious institutions, practices, ideas, and attitudes on the one hand, and scientific institutions, practices, ideas, and attitudes on the other. Without denying the existence of tensions, the primary goal of this series is to explore the vast domain of mutually supportive and/or transformative interactions between scientific institutions, practices, and knowledge and religious institutions, practices, and beliefs. A second goal is to offer the opportunity to make comparisons across space, time, and cultural configuration. The series will cover the entire globe, most major faith traditions, hunter–gatherer societies in Africa and Oceana as well as advanced industrial societies in the West, and the span of time from classical antiquity to the present. Each volume will focus on a particular cultural tradition, faith community, time period, or scientific domain, so that each reader can enter the fascinating story of interactions between science and religion from a familiar perspective. Furthermore, each volume will include not only a substantial narrative or interpretive core, but also a set of primary documents, which will allow the reader to explore relevant evidence, an extensive annotated bibliography to lead the curious to reliable scholarship on the topic, and a chronology of events to help the reader keep track of the sequence of events involved and to relate them to major social and political occurrences.

So far I have used the words "science" and "religion" as if everyone knows and agrees about their meaning and as if they were equally appropriately applied across place and time. Neither of these assumptions is true. Science and religion are modern terms that reflect the way that we in the industrialized West organize our conceptual lives. Even in the modern West, what we mean by science and religion is likely to depend on our political orientation, our scholarly background, and the faith community to which we belong. Thus, for example, Marxists and Socialists tend to focus on the application of natural knowledge as the key element in defining science. According to the British Marxist scholar Benjamin Farrington, "Science is the system of behavior by which man has acquired mastery of his environment. It has its origins in techniques . . . in various activities by which man keeps body and soul together. Its source is experience, its aims, practical, its *only* test, that it works."[7] Many of those who study natural knowledge in pre-industrial societies are also primarily interested in knowledge as it is used and are relatively open regarding the kind of entities posited by the developers of culturally specific natural knowledge systems or "local sciences." Thus, in his *Zapotec Science: Farming and Food in the Northern Sierra of Oaxaca*, Roberto González insists that

Zapotec farmers . . . certainly practice science, as does any society whose members engage in subsistence activities. They hypothesize, they model problems, they experiment, they measure results, and they distribute knowledge among peers and to younger generations. But they typically proceed from markedly different premises—that is, from different conceptual bases—than their counterparts in industrialized societies.[8]

Among the "different premises" is the Zapotec scientists' presumption that unobservable spirit entities play a significant role in natural phenomena.

Those more committed to liberal pluralist society and to what anthropologists like González are inclined to identify as "cosmopolitan science" tend to focus on science as a source of objective or disinterested knowledge, disconnected from its uses. Moreover, they generally reject the positing of unobservable entities, which they characterize as "supernatural." Thus, in an *Amicus Curiae* brief filed in connection with the 1986 Supreme Court case that tested Louisiana's law requiring the teaching of creation science along with evolution, *72 Nobel Laureates, 17 State Academies of Science and Seven Other Scientific Organizations* argued that

Science is devoted to formulating and testing naturalistic explanations for natural phenomena. It is a process for systematically collecting and recording data about the physical world, then categorizing and studying the collected data in an effort to infer the principles of nature that best explain the observed phenomena. Science is not equipped to evaluate supernatural explanations for our observations; without passing judgement on the truth or falsity of supernatural explanations, science leaves their consideration to the domain of religious faith.[9]

No reference whatsoever to uses appears in this definition. And its specific unwillingness to admit speculation regarding supernatural entities into science reflects a society in which cultural specialization has proceeded much further than in the village farming communities of southern Mexico.

In a similar way, secular anthropologists and sociologists are inclined to define the key features of religion in a very different way than members of modern Christian faith communities. Anthropologists and sociologists focus on communal rituals and practices that accompany major collective and individual events: plowing, planting, harvesting, threshing, hunting, preparation for war (or peace), birth, the achievement of manhood or womanhood, marriage (in many cultures), childbirth, and death. Moreover, they tend to see the intensification of social cohesion as the major consequence of religious practices. Many Christians, on the other hand, view the primary goal of their religion as personal salvation, viewing society at best as a supportive structure and at worst as a distraction from their own private spiritual quest.

Thus, science and religion are far from uniformly understood. Moreover, they are modern Western constructs or categories whose applicability to the temporal and spatial "other" must always be justified and must furthermore be interpreted as the means by which we organize our understanding of the actions and beliefs of people who would not have used those terms themselves. Nonetheless it does seem to us not simply permissible but probably necessary to use these categories at the start of any attempt to understand how actors from other times and places interacted with the natural world and with their fellow humans. It may ultimately be possible for historians and anthropologists to understand the practices of persons distant in time and/or space in terms that those persons might use. But that process must begin by likening the actions of others to those that we understand from our own experience, even if the likenesses are inexact and in need of qualification.

The editors of this series have not imposed any particular definition of science or of religion on the authors, expecting that each author will develop either explicit or implicit definitions that are appropriate to their own scholarly approaches and to the topics that they have been assigned to cover.

Richard Olson
Claremont, California

NOTES

1. Tertullian, 1896–1903. "Ad nationes," in *The Anti-Nicene Fathers,* ed. Alexander Roberts and James Donaldson, trans. Peter Holmes (New York: Scribner), 3:133.

2. Christopher Toumey, *God's Own Scientists: Creationists in a Secular World* (New Brunswick, NJ: Rutgers University Press, 1994), 94.

3. Galileo Galilei, *Dialogue Concerning the Two Chief World Systems: Ptolemaic and Copernican* (Berkeley: University of California Press, 1953), 5.

4. James R. Moore, *The Post-Darwinian Controversies: A Study of the Protestant Struggle to Come to Terms with Darwin in Great Britain and America, 1870–1900* (Cambridge: Cambridge University Press, 1979), 60.

5. Stephen Jay Gould, *Rocks of Ages: Science and Religion in the Fullness of Life* (New York: Ballantine, 1999), 4.

6. Ibid., 5.

7. Benjamin Farrington, *Greek Science* (Baltimore: Penguin, 1953).

8. Roberto González, *Zapotec Science: Farming and Food in the Northern Sierra of Oaxaca* (Austin: University of Texas Press, 2001), 3.

9. *72 Nobel Laureates, 17 State Academies of Science and Seven other Scientific Organizations. Amicus Curiae.* Brief in support of Appelles Don Aguilard, et al. v. Edwin Edwards in his official capacity as Governor of Louisiana et al. (1986), 24.

Chronology of Events

600–400 B.C.	Greek pre-Socratic philosophers.
431–404 B.C.	Peloponnesian Wars.
399 B.C.	The trial and execution of Socrates, Plato's teacher.
c. 380 B.C.	Plato founds the Academy, a school in Athens where philosophers and mathematicians exchanged ideas and opinions and sought to advance their disciplines.
335 B.C.	Aristotle, Plato's most famous student, founds his own school, the Lyceum, just outside of Athens, where he brought together philosophers and scientists to pursue research in many fields.
334–323 B.C.	Alexander the Great establishes his empire but dies in Babylon in 323.
300–100 B.C.	Hellenistic period; the great period of Greek scientists, including Euclid, Archimedes, Apollonius, Aristarchus, and many others.
146 B.C.	Romans conquer Greeks.
c. A.D. 30	Jesus executed.

d. c. A.D. 40	Philo of Alexandria (Philo Judaeus): wrote first commentary on creation account in Genesis and originated the handmaiden approach to scripture and faith.
1st century	Two important Latin encyclopedic authors who had a significant influence on the Middle Ages: Seneca (d. 68), who wrote *Natural Questions*, and Pliny (23/24–79), the author of the *Natural History* in thirty-seven books.
2nd century	Claudius Ptolemy wrote the most important books on astronomy and astrology in the ancient world. Galen was the greatest physician in antiquity whose medical treatises were dominant until the seventeenth century.
c. A.D. 184–c. 254	Origen: a famous Christian philosopher and scholar who showed that Greek philosophy was compatible with Christianity.
c. A.D. 204–270	Plotinus: founder of neo-Platonism.
d. c. A.D. 215	Clement of Alexandria: one of earliest Church Fathers to advocate that science and philosophy be studied as handmaidens to theology and faith.
A.D. 313	The Roman emperor Constantine issues the Edict of Milan, or Edict of Toleration, which conferred on Christianity full legal equality with all other religions in the Roman Empire.
A.D. 325	Council of Nicaea, which denounced the Arian heresy and formulated the Nicene Creed proclaiming that the Son was one in being with the Father.
c. A.D. 330–379	Saint Basil of Caesarea: regarded as a saint in the Eastern Orthodox and Roman Catholic Churches, was bishop of Caesarea from 370–379, and an opponent of Arianism. As a famous preacher, many of his homilies were preserved.

A.D. 354–430   Saint Augustine: a prolific Latin author who exerted an enormous influence on medieval theology. Like many church fathers, he advocated the handmaiden approach to secular learning.

A.D. 410   King Alaric's Visigoths sack Rome.

480–524/525   Anicius Manlius Severinus Boethius: called "Last of the Romans, first of the scholastics." He exerted an enormous influence on medieval theology and supplied the basic texts for the disciplines of arithmetic, music, and logic. Boethius was the author or translator of almost all of the numerous treatises that comprised the "old logic," which served Western Europe prior to the introduction of Aristotle's logic in the thirteenth century.

6th century   In 529, the Roman emperor Justinian closes the neo-Platonic School of Philosophy in Athens, which he regarded as a center of paganism. Two of the most important Greek philosophers of this period were: John Philoponus (fl. first half of century), a Christian neo-Platonist, critic of Aristotle, and the author of important commentaries on Aristotle's natural philosophy that subsequently influenced Islamic and Christian authors in the Middle Ages; and Simplicius (c. 500–d. after 533), a commentator on Aristotle's natural philosophy, who defended Aristotle against John Philoponus' criticisms. He, too, had a significant influence on both Islam and the West in the Middle Ages.

c. A.D. 570–632   Muhammad and the beginning of Islam.

c. A.D. 800   Charlemagne crowned Roman emperor by Pope Leo III.

c. A.D. 813–833   Al-Ma'mun, who was Caliph in this period, founded the House of Wisdom in Baghad, which functioned as a research center and a place where translations were made into Arabic.

c. A.D. 870	Vikings discover Iceland.
A.D. 1095	Pope Urban II proclaims the First Crusade to regain the Holy Land.
c. A.D. 1114–1187	Gerard of Cremona: translated more Arabic science and natural philosophy (at least seventy-one works) into Latin than any one else.
A.D. 1126–1198	Averroes (Ibn Rushd): his numerous commentaries on Aristotle's natural philosophy were translated from Arabic to Latin in the thirteenth century and exerted a great influence on Latin scholastic thought.
A.D. 1140	At the instigation of Bernard of Clairvaux, the bishops at the Council of Sens condemned the writings of Peter Abelard.
c. A.D. 1155	Peter Lombard wrote the *Sentences*, in four books, which became the basic textbook in theology for the next 500 years.
c. A.D. 1200	The Universities of Paris and Oxford are in existence.
A.D. 1204	Christian crusaders conquer and sack Constantinople.
A.D. 1210	Aristotle's works, and commentaries on those works, are banned at Paris.
A.D. 1215	Fourth Lateran Council proclaims that God created the world from nothing.
	King John of England signs Magna Carta.
A.D. 1222	Foundation of the University of Padua.
c. A.D. 1250–c. 1295	Major contributions were made to natural philosophy and theology by William of Moerbeke (c. A.D. 1215–c. 1286), the most prolific translator of Greek scientific works into Latin, who translated approximately forty-eight treatises, including major works of Aristotle, Archimedes, and Galen.

Saint Bonaventure (John of Fidanza) (c. A.D. 1221–1274) emphasized the limitations of philosophy and the uncertainty of knowledge; he also presented numerous arguments against the eternity of the world.

Saint Thomas Aquinas (c. A.D. 1224–1274) was not only a great theologian, but was also heavily involved in natural philosophy, writing numerous commentaries on Aristotle's works.

Roger Bacon (c. A.D. 1219–c. 1292) advocated the handmaiden approach to Greek science and philosophy, but also urged experimental studies and the application of mathematics to nature.

c. A.D. 1255	Aristotle's books on natural philosophy are added to the curriculum at the University of Paris.
A.D. 1258	Mongols take Baghdad, the intellectual and spiritual center of the Muslim world.
A.D. 1271	Marco Polo travels to China.
A.D. 1272	Members of arts faculty at University of Paris are required to swear that they will not dispute purely theological questions, but if they find it absolutely necessary, they must resolve all issues in favor of the faith.
A.D. 1277	The bishop of Paris condemns 219 articles, most of which are relevant to natural philosophy and many of which were directed at Aristotle.
A.D. 1309–1377	The Papacy moves from Rome to Avignon, France.
c. A.D. 1307–1321	Dante Alighieri (1265–1321) composed the *Divine Comedy*, consisting of the *Inferno*, *Purgatory*, and *Paradise*.
A.D. 1337	France and England begin the Hundred Years' War.

A.D. 1347   The University of Prague is founded.

A.D. 1347–1349   The Black Death, an epidemic that caused the
                    death of approximately one-third of Europe's
                    population.

A.D. 1372–1377   Nicole Oresme translated four of Aristotle's
                    works from Latin to French at the command of
                    the French king, Charles V. The translations
                    were meant to improve the quality of those who
                    governed France. Oresme was probably the
                    greatest natural philosopher and mathematician
                    of the Middle Ages.

A.D. 1431   Joan of Arc is burned at the stake in Rouen for
                    heresy.

c. A.D. 1450   Johann Gutenberg (c. 1399–1468) invents print-
                    ing from movable metal type, which allows for
                    the replacement of hand-copied manuscripts
                    with printed books, thereby ushering in a revo-
                    lution in the dissemination of learning.

A.D. 1453   Ottoman Turks capture Constantinople and end
                    the Byzantine (Roman) Empire.

                    End of Hundred Years' War between France and
                    England.

A.D. 1492   Columbus discovers the New World.

                    Christians take Granada (Spain) from Muslims.

A.D. 1543   Copernicus' great astronomical treatise, *On the
                    Revolutions of the Heavenly Spheres*, is published.
                    The heliocentric planetary system it espoused
                    would eventually lead to the abandonment of
                    the medieval Aristotelian geocentric worldview.

Chapter 1

Introduction

As the title indicates, this is a book about the history of the relations between science and religion over a period that extends from approximately 400 B.C. to A.D. 1550. Although Greek science and natural philosophy began about two centuries before Aristotle was born, the real beginnings of the dialogue between science and religion commenced with Plato and his student Aristotle (see Figure 1.1). Aristotle's name was chosen for inclusion in the title because his role in the science–religion dialogue that will be described in this volume dwarfs that of Plato. Nicholas Copernicus (A.D. 1473–1543) was chosen to signify the end of this period because his monumental treatise of 1543—*On the Revolutions of the Heavenly Spheres (De revolutionibus orbium coelestium)*—marked the beginning of the end for the medieval worldview (see Figure 1.2). With his proclamation of a heliocentric, or sun-centered, planetary system, Copernicus began the intellectual process that led ultimately to Galileo, Johannes Kepler, and Isaac Newton in the seventeenth century. It was they who vindicated Copernicus and forced the abandonment of the medieval cosmos, a cosmos that had been based on Aristotle's geocentric, or earth-centered, planetary configuration, which had been brought to its fullest development by medieval scholastic natural philosophers and theologians whose opinions and attitudes constitute the substantive content of this book.

Figure 1.1.   Aristotle. From a bust in the Hofmuseum, Vienna.

THE MIDDLE AGES: A TIME OF IGNORANCE AND BARBARISM? OR A PERIOD OF STRIKING INNOVATION?

Scholars in the Renaissance regarded the sixteenth and seventeenth centuries as the beginning of a new age in which Europe was destined

NICOLAVS COPERNICVS
Mathematicus.

*Quid tum? si mihi terra mouetur, Solᵍ, qui escit,
Ac cœlum: constat calculus inde meus.*

M. D. XLI.

Figure 1.2. Nicholas Copernicus. Museum of Torun, Torun, Poland. (Erich Lessing/Art Resource, NY.)

for great achievements. To show that Europe had departed from its dismal past, they divided history into three parts. The first embraced the ancient Greeks and Romans down to the conversion of the Roman emperors to Christianity in the fourth century. The second, the Middle Ages, extended from around 400 to approximately 1400 or 1500—the middle of the three historical ages. The third period was everything after that—a rich, positive period that came to be called the Renaissance and was the modern world of the authors who named it. The Renaissance revived the Greek and Roman classics in literature, history, and science and turned away from the scholastic literature of the Middle Ages. It was regarded as a fertile period in the arts and sciences. Renaissance scholars prided themselves on a progressive outlook that went beyond the achievements of the ancients.

The West European Middle Ages extending from around 500 to 1500, have been, and often still are, perceived as a sterile, superstitious period in the history of Western civilization. A distinguished medieval historian encapsulated a common view of the Middle Ages when he reported that

The Middle Ages were condemned as "a thousand years without a bath" by one well-scrubbed nineteenth-century writer. To others they were simply "the Dark Ages"—recently described (facetiously) as the "one enormous hiccup in human progress." At length, sometime in the fifteenth century the darkness is supposed to have lifted. Europe awakened, bathed, and began thinking and creating again. After a long medieval intermission, the Grand March resumed. (Hollister 1994, 1)

It is difficult to imagine a more inaccurate and misleading assessment. Yet the attitudes described above were commonplace between the seventeenth and nineteenth centuries. In the eighteenth century, for example, Voltaire, the famous French author and philosopher, spoke for many when he referred to "the *history of the Middle Ages*" as "a barbarous history of barbarous peoples, who became Christians but did not become better because of it" and also that "it is necessary to know the history of that age only in order to scorn it" (Grant 2001, 332). Such attitudes were commonplace in the nineteenth and twentieth centuries. As recently as 1992, a reputable historian said of the Middle Ages:

In all that time nothing of real consequence had either improved or declined. Except for the introduction of waterwheels in the 800s and windmills in the late 1100s, there had been no inventions of significance. No startling new ideas had appeared, no new territories outside Europe had been explored. Everything was as it had been for as long as the oldest European could remember. (Manchester 1992, 27)

Unaware of the coming of the Renaissance in the sixteenth century, medieval society was totally unprepared for the advent of a new age, believing that the future would be just like the previous thousand years of darkness. This is the Rip Van Winkle view of the Middle Ages: a thousand years of sleep before the Renaissance produced the great awakening of Western Europe. The relentless attacks and harsh criticisms of the Middle Ages have penetrated popular culture to the extent that the Middle Ages are regarded as an age of superstition, cruelty, and ignorance. If journalists, for example, wish to describe cruelty, they frequently achieve this by introducing the term "medieval," as when a reporter had occasion to mention that, "the Nazis combined the worst of medieval cruelty with twentieth-century technology and created a terrifying synthesis" (Reid 1998).

Perhaps the most powerful illustration of bias against the Middle Ages concerns Christopher Columbus' voyage of discovery to the New World in 1492. Many came to believe that the most significant achievement of Columbus' voyage was the discovery that the earth is not flat—as was universally believed in the Middle Ages—but round. This is utterly false. No educated person in the Middle Ages believed in a flat earth (Russell 1991). They all knew it was round. Their authority was Aristotle. In his major cosmological treatise, *On the Heavens*, Aristotle emphatically declared the earth a sphere and even presented an estimate of its circumference. All who were educated in the universities of the Middle Ages would have read that passage. But it could be found in many other treatises they might also have read. No one would have doubted it. And yet, nineteenth-century authors were able to construct a falsehood still widely believed that everyone in the Middle Ages believed in a flat earth until Columbus' voyage proved its sphericity.

Despite the powerful overall general bias against the Middle Ages, only one aspect of that hostility will concern us in this volume—the assault against the medieval study of logic, natural philosophy, and theology at the approximately sixty universities created in the period from 1200 to 1500, an assault that began in the sixteenth century and continued on to the end of the nineteenth century. The kind of learning signified by these disciplines was collectively known as scholasticism, and its practitioners were called *scholastics*. In logic and natural philosophy, medieval scholastics were rightly regarded as followers of Aristotle because their primary activity was to study and comment upon Aristotle's logic and natural philosophy. Until the late fifteenth century, when many of the works of Plato were translated from Greek into Latin and previously unknown works of other Greek authors were also translated, Aristotle's works had gone unchallenged. For two and a half centuries—from around 1200 to 1450—Aristotle's numerous works were dominant in medieval university education. Indeed, there were no rivals. But beginning around 1450, Aristotle would have many rivals as numerous previously unknown works were translated and new schools of thought developed to challenge his dominance. As rival philosophies emerged, an intellectual struggle developed in which there were strong mutual criticisms. As the scholastic system of education weakened in the seventeenth century, more and more critics of that system voiced their objections to it and attacked it by serious analysis, but more often by lampooning the way scholastics thought and argued. The cumulative impact of these criticisms played a crucial role in creating the idea that the Middle Ages were an intellectually sterile period.

The attack on medieval scholasticism began in the Middle Ages with Francesco Petrarch (A.D. 1304–1374), who admired Aristotle but criticized his followers, because they made doing logic an end in itself instead of a path to something else. In the fifteenth and sixteenth centuries, the criticisms were directed overwhelmingly against scholastic logic and theology, although there were attacks directed against Aristotelians in general. In the fifteenth century, Lorenzo Valla (A.D. 1407–1457) attacked Aristotelians by falsely claiming they took an oath never to contradict Aristotle. Valla emphasized the narrow-mindedness of scholastic Aristotelians, declaring: "modern Peripatetics [that is, Aristotelians] are intolerable. They deny a person who does not adhere to any school the right to disagree with Aristotle" (Rummel 1995, 160). The criticisms intensified in the sixteenth century when famous figures such as Desiderius Erasmus, Thomas More, Juan Luis Vives, and others attacked scholastic logic and theology.

Erasmus used the formidable weapon of ridicule to attack scholastic logicians and theologians. By the end of the Middle Ages, scholastic logic had become virtually unintelligible to anyone not immersed in its strange juxtapositions of words and the bizarre sentences it used as examples. Theology had become highly analytical, and it used logic and natural philosophy extensively, as we shall see. Theologians were into abstract and strange questions about whether God could do this or that action. In his famous treatise, *In Praise of Folly*, Erasmus cited such questions to show how absurd theology had become. He presented the following sequence of questions, which he claimed contemporary theologians discussed: "What was the exact moment of divine generation? Are there several filiations in Christ? Is it a possible proposition that God the Father could hate his Son? Could God have taken on the form of a woman, a devil, a donkey, a gourd, or a flintstone? If so, how could a gourd have preached sermons, performed miracles, and been nailed to the cross?" (Erasmus 1993, 87). Such strange questions prompted Erasmus to say that theologians were "so busy night and day with these enjoyable tomfooleries, that they haven't even a spare moment in which to read the Gospel or the letters of Paul even once through" (Erasmus 1993, 93). Erasmus, as did many others, felt that theology had been transformed into an incomprehensible discipline that was intelligible only to professional theologians. It had nothing to do with traditional religion, largely ignoring the Gospels and the Bible.

The most telling criticism against scholastic logic was delivered by Juan Luis Vives (A.D. 1493–1540), a Spaniard born in Valencia. He was trained in scholastic logic while in Paris but abandoned it early on to embrace the new humanism that was sweeping Europe. In 1520, Vives published *Against the Pseudodialecticians*, in which he used his knowledge of scholastic logic to attack it. Vives' favorite tactic was to show how scholastic logicians had developed an extensive and intricate technical jargon that was unintelligible to all but themselves. They also used the word order of a sentence or proposition to show how that could drastically change the meaning of the sentence. In the Latin language, word order does not affect the meaning of a sentence. But scholastic logicians changed that by using word order to convey different meanings. Changes in word order could be crucial to a logician but meaningless to a non-logician. It is important to realize that historians of logic find the medieval practice of changing word order to convey different meanings with the same words to be rational and an early version of formal logical notation. But it was simply unintelligible to non-logicians and was easily made the object of ridicule. In referring to the logicians, Vives declared that

they have invented for themselves certain meanings of words contrary to all civilized custom and usage, so that they may seem to have won their argument when they are not understood.

For when they are understood, it is apparent to everyone that nothing could be more pointless, nothing more irrational. So, when their opponent has been confused by strange and unusual meanings and word-order, by wondrous suppositions, wondrous ampliations, restrictions, appellations, they then decree for themselves, with no public decision or sentence, a triumph over an adversary not conquered but confused by new feats of verbal legerdemain. Truly, would not Cato, Cicero, Sallust, Livy, Quintilian, Pliny, and Marcus Varro (recognized as the first Latin writers on logic) be utterly at a loss to hear one of these sophisters make statements like these:

> When he is full of drink, swear on the stone Jove that he has not drunk wine, because he has not drunk wine that is in India.
> When he sees the King of France attended by a great retinue of servants, say This King does not have servants because he does not have those who wait on the King of Spain.
> Though Varro is a man, yet he is not a man because Cicero is not himself Varro.
> That a head no man has even though no man lacks a head.
> There are more non-Romans than Romans in this hall, in which there are a thousand Romans and two Spaniards. . . .
> Socrates and this donkey are brothers.

Two contradictory statements are also in a contradictory sense true. (Guerlac 1979, 57–61)

Vives' treatise was brief and widely read. It had a great impact on sixteenth-century thought. Vives was encouraged to write his treatise after he read a letter by Sir Thomas More to one Martin Dorp, written in 1515. In this letter, More spoke of a logic textbook called the *Little Logicals*. Boys learning their logic from this book, said More, were taught a series of piddling rules and many falsehoods. The boys were

taught to distinguish among statements of this kind, "The lion than an animal is stronger" and "The lion is stronger than an animal"—as if they did not mean the same thing. Actually both statements are so clumsy that neither of them means much of anything, but if they do mean anything they doubtless mean the same. And these differ just as much, "Wine twice I have drunk" and "Twice wine I have drunk"; that is to say, a lot according to those logickers but in reality not at all. Now if a man has eaten meat roasted to burning, they want him to be speaking the truth if he puts it this way "I raw meat have eaten," but not if he says "I have eaten raw meat." (Guerlac 1979, 171–173)

Thus did More ridicule and reject the idea of using the word order of sentences to convey different meanings. He was also critical of theologians who were absorbed in the pursuit of trivial questions "that are empty in themselves and useless to men who are empty of all the rest of knowledge" (Guerlac 1979, 191). These theologians could not discuss scripture properly because they were ignorant of the writings of the ancients and deficient in the Latin language.

There were other critics of scholastic logic and theology, but only one more will be mentioned: Martin Luther. In 1517, Luther wrote a treatise that reveals its intent in the title, *Against Scholastic Theology*. Luther, who had studied theology, was disturbed by the rationalistic theology that still dominated the theological schools of his day. He formulated ninety-seven criticisms against the theologians, among which are the following:

43. It is an error to say that no man can become a theologian without Aristotle. This in opposition to common opinion.
44. Indeed, no one can become a theologian unless he becomes one without Aristotle.
45. To state that a theologian who is not a logician is a monstrous heretic— this is a monstrous and heretical statement. This in opposition to common opinion.
46. In vain does one fashion a logic of faith, a substitution brought about without regard for limit and measure. This in opposition to the new dialecticians.
47. No syllogistic form is valid when applied to divine terms. This in opposition to the Cardinal.
48. Nevertheless it does not for that reason follow that the truth of the doctrine of the Trinity contradicts syllogistic forms. This in opposition to the same new dialecticians and to the Cardinal.
49. If a syllogistic form of reasoning holds in divine matters, then the doctrine of the Trinity is demonstrable and not the object of faith.
50. Briefly the whole Aristotle is to theology as darkness is to light. This in opposition to the scholastics. (Luther 1957, 12)

Luther obviously condemned (Luther 1957, 12) the use of logic in theology and faith, but was equally repelled by the use of Aristotle's works in theology.

The criticisms of scholastic logic and theology continued on into the seventeenth and eighteenth centuries. As time passed, fewer and fewer of the critics of scholastic theology and logic had any genuine acquaintance with those disciplines as they were understood in the Middle Ages. Nonetheless, it became traditional and customary for those ignorant of medieval and early modern Aristotelian scholastic thought to harshly criticize it. English philosophers, such as Francis Bacon, John Locke, and Thomas Hobbes, who wrote in the seventeenth century, were severe critics of scholastic logic and theology. Denunciations of scholasticism continued on into the eighteenth and nineteenth centuries.

Medieval natural philosophy was subjected to a somewhat different fate than logic and theology. It was not the object of severe criticism until the seventeenth century. The assault on natural philosophy by seventeenth-century authors such as Galileo, Francis Bacon, Thomas Hobbes, Pierre Gassendi, and René Descartes proved fatal to that discipline, which lost all credibility. By the end of the century, Aristotelian natural philosophy was no longer a major intellectual force. The greatest blow against scholastic natural philosophy was delivered by Galileo Galilei (A.D. 1564–1642), who used wit, sarcasm, and irony as basic weapons against these philosophers. He never tired of emphasizing the slavish devotion of Aristotle's followers to the master.

In one of his most significant works, the *Dialogue Concerning the Two Chief World Systems—Ptolemaic and Copernican*, Galileo presented striking examples of unreasonable devotion to Aristotle in the face of evidence that obviously contradicted Aristotle's interpretation. Galileo related an incident of his attendance at an anatomical dissection in which the anatomist demonstrated that the nerves originate in the brain and not in the heart, as Aristotle believed. The anatomist turned to an Aristotelian natural philosopher and asked him "whether he was at last satisfied and convinced that the nerves originated in the brain and not in the heart. The philosopher, after considering for awhile, answered: 'You have made me see this matter so plainly and palpably that if Aristotle's text were not contrary to it, stating clearly that the nerves originate in the heart, I should be forced to admit it to be true'" (Galileo 1962, 108). In a similar vein, Galileo mentioned telescopic observations that he had made with the telescope, which was invented in 1608. Galileo was convinced that Aristotelian natural philosophers did not accept the validity of telescopic observations. To illustrate this, Galileo has Simplicio, who represents the Aristotelian natural philosopher in the *Dialogue*, explain why he would not read books describing telescopic observations. Simplicio declares:

Frankly, I had no interest in reading those books, nor up till now have I put any faith in the newly introduced optical device. Instead, following in the footsteps of other Peripatetic philosophers of my group, I have considered as fallacies and deceptions of the lenses those things which other people have admired as stupendous achievements. (Galileo 1962, 336)

Galileo had great respect for Aristotle, but only contempt for the scholastic natural philosophers who, he was convinced, mindlessly followed him. More than anyone else, Galileo shaped the judgments about medieval scholasticism for the centuries that followed. His works were widely read, and his reputation was enormous. And yet one caution should be kept in mind: Galileo's attack on scholasticism was leveled against the scholastics of his time, in the dying days of medieval scholasticism. During the thirteenth to fifteenth centuries, however, medieval natural philosophers abandoned Aristotle's interpretations on numerous occasions. Unfortunately, Galileo's criticisms of contemporary Aristotelians as slavish followers of Aristotle without any significant thoughts of their own was extended in blanket fashion to all scholastic natural philosophers as far back as the thirteenth century. Thus was the whole of the Middle Ages condemned as a sterile period in the intellectual life of Western Europe.

The nineteenth and twentieth centuries produced more of the same attitude. For example, Francis Bowen, a professor of natural religion and moral philosophy at Harvard College, wrote in his history of philosophy that "Aristotelic premises were evoked to support theological conclusions. Novelty was shunned, because it immediately incurred suspicion of heresy" (Bowen 1885, 2). Many expounded the ludicrous idea that any departure from Aristotle's thought was regarded as heresy, as did Harald Höffding in his *History of Modern Philosophy*, first published in 1900. A remark by Charles Singer, a well-known historian of science and medicine, will conclude this skeletal survey of anti-scholastic hostility. In 1941, Singer published a brief book on the history of science in which he found occasion to declare that "many attempts have been made to rehabilitate the intellectual achievement of the Middle Ages. So far as science is concerned they have been unsuccessful. There is no reason to reverse the decision that in this domain the period is one of intellectual degradation" (Singer 1941, 161).

Enough has now been said to convey a picture of traditional, negative attitudes toward the Middle Ages. Beginning in the late nineteenth century, and continuing through the twentieth, medieval historians—in contrast to those cited in the preceding paragraph, and many others like them, who were largely ignorant of the Middle Ages—began to present a more accurate sense of medieval history in all its manifestations. In the history of science, Pierre Duhem was a pioneer in the study of medieval science and natural philosophy. His ten-volume work, *Le Système du Monde* (published from 1913 to 1959), covered the period from Plato to Copernicus but was mostly on the Middle Ages. Many other scholars have followed in his footsteps, producing a detailed picture of medieval science and natural philosophy. What have the extensive research efforts on all aspects of the Middle Ages produced? How have they altered the way we view the Middle Ages?

Contrary to Traditional Misconceptions, the Middle Ages Was an Innovative Period

Despite all the anti-medieval passages cited to this point, the Middle Ages was one of the most innovative periods in human history. Significant advances were made in commerce and numerous other fields. Among the innovations in technology were eyeglasses, the magnetic compass, the mechanical clock, firearms and the cannon, ship rudders, cranks to convert continuous rotary motion to reciprocating motion, and the printing press. Higher education saw the creation of the university. For the first time in banking, there were bills of exchange, checks, and marine insurance. These advances in commerce were supplemented by the development of codes of maritime law and the formation of joint stock companies. In medicine, human cadavers were dissected for the first time for teaching purposes in medical schools. Government changed with the rise of the nation state and the creation of the Magna Carta as well as the English parliament, the first representative government. The Middle Ages furthermore laid the basis for the modern corporation, and in law, the foundation for the Western legal system. Polyphonic music is a product of the Middle Ages, and it was during this period that the Arabic number system was first adopted by the West. Medieval explorers expanded the horizons of Europe as never before. The Vikings reached the shores of Newfoundland around 1000. Before 1500, European explorers Bartholomew Diaz and Vasco da Gama reached India by rounding the Cape of Good Hope, followed a few years later by Christopher

Columbus, who reached the New World and began the long period of European imperialism and colonization. Finally, and most relevant for this volume, the Middle Ages, for the first time in the history of civilization, became a society in which innumerable questions about nature were raised and then resolved almost exclusively by the use of reason. That extraordinary achievement laid the foundation for the advancement of science, which depends upon reasoned analysis.

Taken collectively, these are extraordinary achievements. Many other contributions could be cited, but enough have been mentioned to show that the Middle Ages were a fertile and inventive period during which the foundations of Western civilization were laid and the way prepared for uninterrupted advancements over the next 500 years. These significant achievements occurred within a Christian civilization, although most had little to do with the religious aspects of Christianity. One great exception concerns the philosophical disciplines of logic and natural philosophy, which were derived from the pagan Greeks. Natural philosophy contained a different account of the world than was found in the Bible. From its very beginnings, Christianity found this a problem that had to be resolved. How should Christians view philosophical works and philosophical discussions about the world that might impinge upon Christian belief? The issue of the relations of science and Christianity derive from the interplay of doctrines of the Christian religion with pagan science, however one may view that science.

RELIGION AND SCIENCE AMONG THE GREEKS PRIOR TO THE EMERGENCE OF CHRISTIANITY

In the most general sense, this volume is a study of the relations between reason and revelation as embodied, respectively, in the tradition of Greek philosophical thought and Christianity. It is the relationship characteristic of Greek learning acquired by the natural light of human reason, which stands in contrast to revelation—that is, religious truths revealed by God and accepted on the basis of the Christian faith. A number of church fathers, especially Saint Augustine, emphasized the distinction between reason and revelation. Nevertheless, they believed in one truth. A truth arrived at by reason was as good as a truth given by revelation. Thus, there was a basic agreement between natural and revealed knowledge. There are, however, certain revealed truths that are beyond reason and known only by faith. Certain revealed truths can be known by human reason, but others cannot. The existence of God, for example, is a revealed truth that can also be known by reason; that is, the existence of God is thought by many to be capable of proof by reason alone, independently of revelation. Revealed truths that cannot be known by reason are the Trinity and Incarnation, which can only be objects of faith.

Although there were other radically different viewpoints, as we shall see, the position just described encouraged Christians to believe that pagan Greek philosophy and natural philosophy were useful rational tools to aid in the understanding of scripture. One of the greatest and most influential Christian authors was Saint Augustine, who insisted that revelation was superior to reason. As one historian of philosophy has expressed it: "Augustine was never to forget that the safest way to reach truth is not the one that starts from reason and then goes on from rational certitude to faith, but, on the contrary, the way whose starting point is faith and then goes on from Revelation to reason" (Gilson 1938, 17).

Relations between science and religion in the early Christian world of the late Roman Empire were really concrete instances of the broader relationship involving reason and revelation. Few Christians in late antiquity actually did anything that we might call science, but most found they could not comment on the creation account in Genesis without introducing natural philosophy into their accounts (see the discussion of hexameral literature in chapter four). In this volume, I shall emphasize natural philosophy, and to a much lesser extent, medicine. These were the disciplines that affected, and were affected by, religion. I shall have relatively little to say about the exact sciences of technical astronomy, optics, and statics, which were largely noncontroversial and did not stir the passions of theologians. Although I shall use the expression "science and religion," by science, I usually mean the discipline of natural philosophy, and by religion, the discipline of theology.

The long interrelationship between science and religion, with which most college- and university-educated individuals have at least some familiarity, cannot be traced back to the beginnings of civilization. Its true beginnings, as we shall see, occurred with the advent of Christianity in the Roman Empire. This does not mean that there was no connection between science and pagan religion prior to Christianity. But the relationship was radically different from what was to come.

The Role of Religion in Greek Science

Not long after the beginnings of science and natural philosophy in Greece, the first known clash between science and religion in the pre-Christian Greek world occurred, producing the first known victim of religious persecution. In the time of Pericles, Anaxagoras of Clazomenae (c. 500–428 B.C.), the last of the Ionian pre-Socratic natural philosophers and a friend of Pericles, was apparently persecuted for impiety because he believed the sun was a mass of red-hot metal and therefore, presumably not a divine celestial object. The charge of impiety was probably brought by Pericles' enemies, who apparently saw a good opportunity to attack him, using the pretext of his friendship with the atheistic natural philosopher Anaxagoras. This resulted in the banishment of Anaxagoras from Athens. According to Diogenes Laertius (fl. early third century A.D.), Anaxagoras committed suicide, as we learn from this epigram:

> The sun's a molten mass,
>   Quoth Anaxagoras;
> This is his crime, his life must pay the price.
>   Pericles from that fate
>   Rescued his friend too late;
> His spirit crushed, by his own hand he dies. (Diogenes 1950, 1:145)

What all this reveals for the relations between science and religion is that the Greeks of Anaxagoras' time believed that the celestial region was divine, and they therefore found reason to persecute Anaxagoras when he dared proclaim the sun a mass of red-hot metal.

Another clash between science and religion occurred in the third century B.C., when Aristarchus of Samos became the first to proclaim that the cosmos is really heliocentric rather than geocentric. He displaced the earth as center of the world with the sun, and then set the earth moving around the sun with an annual motion while simultaneously rotating daily on its axis. In reaction to this revolutionary move, Cleanthes the Stoic (263–232 B.C.), the second head of the Stoic school, is reported to have charged Aristarchus with impiety, because he removed the "hearth of the universe" from the center of the world and set it into motion. Nothing happened to Aristarchus, and no such charge was ever brought officially by any religious or governmental body. To my knowledge, no similar case arose in the Greek world prior to the Christian era. It is noteworthy that these two instances were both relevant to the physical structure of the universe, that is, to cosmology.

More significant than these two rather isolated incidents involving science and religion, but also concerned with the physical cosmos, are those relations that involved the attitudes of scientists and natural philosophers toward divinity and the gods. The two greatest philosophers of ancient Greece, Plato and Aristotle, who were both involved with science and natural philosophy, had the most to say about the interrelations of science and religion.

In his most scientific treatise, the *Timaeus*, Plato may have been the first to assume that the world was created by a god, or Demiurge, from disorderly, chaotic pre-existent matter using an eternal, non-physical model of living creatures to fashion all living things. Plato's Demiurge probably represents a divine reason that makes a physical world from unwieldy, uncreated matter. The Demiurge is not omnipotent but seeks to fashion an orderly, rational world from matter that is recalcitrant and difficult to shape. In the *Timaeus*, Plato sought to show the world as an entity guided by reason. In his account, the Demiurge makes the world as close as possible to the eternal model. Although the physical world is but a copy of the model, it is the best possible world that could be created.

Timaeus explains that the Demiurge created the world because he is good "and in the good no jealousy in any matter can ever arise." Therefore,

being without jealousy, he desired that all things should come as near as possible to being like himself. . . . Desiring, then, that all things should be good and, so far as might be, nothing imperfect, the god took over all that is visible—not at rest, but in discordant and unordered motion—and brought it from disorder into order, since he judged that order was in every way the better. (Plato 1957, 33)

Plato used most of the *Timaeus* to describe the natural phenomena of our world, discussing a wide variety of topics, including the formation of the four elements and how numerous creatures of the world function. Indeed, he included much that is physical and biological. In all this, Plato was usually teleological; that is, he wished to show that there is design in nature. Whether there is one god or many gods was of no concern to Plato. The issue of monotheism or polytheism did not arise.

Aristotle took a radically different path than Plato. Aristotle arrived at his notion of God by a process of reasoning that involved causal action. For Aristotle, change was a form of motion. Every motion in the universe requires an external cause or agent. All bodies in the universe are subject to change, because they possess matter, which is always in the process of changing. Aristotle regarded change and motion as inferior to something that might be incorruptible and changeless. Was there such an entity? Yes. Aristotle was convinced that the circular motions of the celestial bodies and all changes below the moon and on the earth—that is, all change in the universe—were ultimately traceable to a single cause, which was itself unchangeable and immobile and, therefore, perfect. In Aristotle's system, this was God, which he called the Unmoved Mover, or Prime Mover. Aristotle's God is eternal and incorporeal. He is located at the outermost celestial sphere, the sphere of the fixed stars, for as Aristotle explains in discussing different senses of the term "heaven," "we recognize habitually a special right to the name 'heaven' in the extremity or upper region, which we take to be the seat of all that is divine" (Aristotle, *On the Heavens* 1984, 1:462). Unlike Plato's creator God, Aristotle's God did not create the world and has nothing to do with it. Although all things move in order to get as close as possible to God, the Unmoved Mover, the latter has no knowledge of their existence. In Aristotle's view, the motion of physical bodies in the universe is caused by contact between a moving agent and the body being moved—that is, contact between mover and moved. But the incorporeal, Unmoved Mover does not cause motion in this way. It causes motion in a non-physical manner by being an object of desire. God is loved by the outermost sphere of the fixed stars as it moves round and round to get as close as possible to the Unmoved Mover. Thus, the Unmoved Mover acts as a final cause because, with no effort or activity, it causes the celestial orb with which it is associated to seek it as an object of love. Indeed, each celestial sphere below the outermost sphere of the fixed stars—Aristotle assumed fifty-four of them to make a grand total of fifty-five—had its own incorporeal unmoved mover, which was often called an intelligence. Of the fifty-five unmoved movers, the first one, which is farthest removed from the earth and is associated with the sphere of the fixed stars, was regarded as the first among equals. Each of the fifty-five spheres moves because it loves its unmoved mover, a concept that gave rise to poetic utterances that in one way or another proclaimed " 'tis love that makes the world go round." Motions in the celestial region, by mechanisms that Aristotle failed to describe, cause all the changes and movements that occur on earth and its environs.

Since Aristotle's God did not create the world and has nothing to do with it, what does Aristotle's God do? He engages in the highest activity possible, namely thought. But what does he think about? He cannot think about anything outside of himself, because all such things are changeable and therefore imperfect and necessarily inferior. The Unmoved Mover must think only about changeless things. Hence he thinks only about himself, because only he is worthy of his own thoughts.

For Aristotle, as for most Greek natural philosophers, the celestial region was divine, composed of a special incorruptible ethereal substance that suffered no change other than of position as it moved eternally with a natural circular motion. Thus, Aristotle's cosmos had important divine characteristics. It was teleological, because Aristotle believed that all things exist for an end or purpose and that all things strive to realize their full potential. But just as important, Aristotle had a sense of awe and wonder about the universe that was certainly religious in tone.

Claudius Ptolemy (c. A.D. 100–c. 170) was the greatest astronomer and astrologer of antiquity. His lengthy and monumental treatise the *Mathematical Syntaxis*, which is known universally by its Arabic title, *Almagest*, was the fundamental astronomical work used by all astronomers until the sixteenth century. His work on astrology, the *Tetrabiblos*, or "four-parted book," was easily the most widely used treatise on astrology during the Middle Ages in both Islam and the Latin West. Indeed, it is still in use today. Ptolemy followed Aristotle in dividing theoretical knowledge into theology (or metaphysics), mathematics, and physics. However, where, Aristotle regarded theology as the most important and highest branch of theoretical knowledge, Ptolemy assigned that role to mathematics. Ptolemy asserted that theology and physics "should rather be called guesswork than knowledge, theology because of its completely invisible and ungraspable nature, physics because of the unstable and unclear nature of matter." He concluded that philosophers will never agree on these two disciplines. But mathematics is radically different: "Only mathematics can provide sure and unshakeable knowledge to its devotees, provided one approaches it rigorously. For its kind of proof proceeds by indisputable methods, namely arithmetic and geometry. Hence we were drawn to the investigation of that part of theoretical philosophy, as far as we were able to the whole of it, but especially to the theory concerning divine and heavenly things. For that alone is devoted to the investigation of the eternally unchanging." In Ptolemy's judgment, astronomy is a branch of mathematics. It makes men see clearly and, moreover, "from the constancy, order, symmetry and calm which are associated with the divine, it makes its followers lovers of this divine beauty, accustoming them and reforming their natures, as it were, to a similar spiritual state" (Ptolemy 1984, 36, 37). The superiority of mathematics, and therefore astronomy, over theology and physics is made apparent when Ptolemy explains how mathematics can help theology and physics.

From all this, it seems plausible to infer that for Ptolemy doing mathematics or astronomy was akin to a religious experience. It was the contemplation of the eternal and unchanging divine celestial region that attracted Ptolemy and gave his astronomical research ethical dimensions. Ptolemy did not, however, reveal any further interest in religion, such as in prayers, or rituals. He was only interested in acknowledging the divine in the unchanging aspects of the cosmos.

What is striking about the attitudes of Plato, Aristotle, Ptolemy, and other Greek natural philosophers is that they did not worship the divinities whose existence they assumed. The gods and divinities they described in their physical treatises were abstract and remote. Moreover, there was no theology that analyzed or studied sacred texts, because there were no significant scriptural texts. Aristotle composed an enormously important treatise on metaphysics, or theology, as he called it, which later came to serve Christians in their difficult analyses of sacred scripture. In the pagan world, Aristotle's treatise on metaphysics dealt with unchanging, incorporeal, and separate substances, the highest kinds of beings in the universe. But these beings, often called intelligences, were little more than abstractions. Aristotle's metaphysics did not become a dominant analytical tool until it was applied to the Christian God in late antiquity.

Science and Natural Philosophy in the Ancient and Medieval Periods

G.E.R. Lloyd rightly explained that "science is a modern category, not an ancient one. There is no one term, in Greek or Latin, that is exactly equivalent to our 'science'." Lloyd mentioned a number of Greek terms "in which the ancients themselves describe what we should call their scientific work." They are terms that may be rendered in English as "inquiry concerning nature," "love of wisdom," "philosophy," "knowledge," and "speculation." "Thus," Lloyd continued, "a good deal of what we know as early Greek science is embedded in philosophy, and this remains true, though to a lesser degree, of the period after Aristotle" (Lloyd 1973, xiii).

Just because the Greeks did not have a specific term, or terms, for "science" does not, of course, mean that they were not doing science in a manner that would be recognized as such by modern scientists. If our goal is to describe the relations between science and religion from early Christianity to the end of the Middle Ages, we must have a reasonably good idea of what science was to those who were engaged in such activities. There are those who believe that modern science differs so radically from the science of late antiquity and the Middle Ages that they would deny altogether that science existed in these early periods. What, then, were they doing? The reply would probably be natural philosophy, which was concerned with all aspects of motion and change in the physical world, but was not itself a science. It seems essential, then, to know what science and natural philosophy were and, if science did exist, how, if at all, was it related to natural philosophy?

In the chapter on Aristotle, I shall describe his division of knowledge and show how he related metaphysics (or theology), mathematics, and natural philosophy. Here we need only mention that Aristotle distinguished the mathematical sciences—astronomy, optics, and mechanics—from natural philosophy. The mathematical sciences—or, as we would call them, the exact sciences—fell midway between natural philosophy and mathematics and were therefore regarded as middle sciences by Aristotle and his medieval followers, who used the Latin phrase *mediae scientiae*—literally "middle sciences." They were also called "mathematical sciences" (*scientiae mathematicae*). Aristotle recognized that the middle sciences involved the application of mathematics to natural phenomena. But he believed the middle sciences belonged more to mathematics than natural philosophy, a theme that was discussed further in the Middle Ages. Those who wish to deny the real existence of science in the Middle Ages would do well to remember that astronomy in the Middle Ages was regarded as a science and actually called a science, albeit a middle, or mathematical science. Moreover, it was usually known by the Latin term *astronomia*, from which, of course, we derive our modern name for the science of astronomy. The practitioners of medieval astronomy were not, however, called astronomers.

If the exact sciences such as astronomy and geometrical optics were not categorized as belonging to natural philosophy, what did belong to that discipline? All phenomena in the heavens and on earth, both animate and inanimate, that moved and changed were regarded as the legitimate province of natural philosophy. This included phenomena we would today assign to various modern sciences, such as geology, meteorology, physics, cosmology, chemistry, biology, and others. When scholars in the ancient and medieval periods discussed earthquakes and other motions of the earth, those would be included today in the discipline of geology; motions of the tides would be considered oceanography; how the four elements (earth, water, air, and fire) become different compounds and how those compounds change into other things would belong to chemistry. Many other examples could be given, but these should suffice to convey the idea that natural philosophers considered problems that belong to many modern sciences, even though neither the names of these modern sciences, nor the sciences themselves, existed in the ancient and medieval periods. Only the phenomena did, many of which were identified and discussed in a variety of contexts, usually within the framework of a treatise on natural philosophy.

We may conclude from this that there were a few recognized sciences in the period with which we are concerned. During the late Middle Ages, following Aristotle, sciences, such as astronomy and optics, were called middle sciences, or mathematical sciences. The term mediae scientiae, as we saw, was used to identify these few mathematical sciences. Thus, the medieval Latin term *scientia* was used for mathematical astronomy, geometric optics, music or harmonics, and mechanics, especially statics, which was known as the "science of weights" in the Middle Ages. When these disciplines are discussed, we are justified in translating *scientia* as *science*. Contrary to a commonly held opinion, the word science was not first used in the nineteenth century but was first employed, in a limited sense and in its Latin form in the late Middle Ages.

As we just saw, the subject matters of many other modern sciences were immersed in the literature of natural philosophy but not identified with any science. Bits and pieces of many sciences were discussed in the literature of natural philosophy over the 1,500 years with which we are concerned. In the seventeenth to nineteenth centuries, many sciences came into being with their modern names. Some, if not many, of them, first appeared in the literature of natural philosophy during the ancient and medieval periods. It is, therefore, quite appropriate to regard natural philosophy as the mother of all sciences.

Natural philosophy considered many questions relevant to one modern science or another and also considered many questions that bear no relationship to any modern science. For example, whether the whole earth is habitable and whether the earth is spherical are questions relevant to geography. Some questions properly belong to astronomy and cosmology including, for example, (Grant 1974, 199–210)

whether spots appearing in the moon arise from differences in parts of the moon or from something external;

whether the mass of the whole earth—that is, its quantity or magnitude—is much less than certain stars;

whether a comet is of a celestial nature or [whether it is] of an elementary nature, say of a fiery exhalation;

whether it is possible that there are several worlds (Grant 1974, 204, 205, 207).

Other questions are pertinent to physics, including

whether the existence of a vacuum is possible;

whether, if a vacuum did exist, a heavy body could move in it;

whether, in local motion, velocity is measured according to distance traversed;

we inquire what it is that moves a projected body upwards after separation from what has projected it;

whether motion could be accelerated to infinity (Grant 1974, 201–203).

There were also questions on optics, a branch of physics, as we see in these questions:

whether every visual ray is refracted in meeting a denser or rarer medium;

whether every visual ray is reflected when it meets a denser medium;

whether a halo could be produced by refraction of visual rays (Grant 1974, 208).

Many questions are not relatable to any particular science or sciences, including, for example, whether the generation of one [thing] is the corruption of another;

whether augmentation is generation;

whether that which is increased in augmentation remains the same before and after;

whether every corruptible thing has a definite period of duration (Grant 1974, 206).

It is obvious that the mother of all sciences was never itself regarded as a science but was, nonetheless, a storehouse of discussions on scientific themes and problems. It was, as we shall see, also a repository for hypothetical questions that compelled natural philosophers to cope with imaginary conditions that required the application of Aristotelian ideas to concepts and ideas that were alien to Aristotle. From these sources of knowledge, natural philosophy, not the exact mathematical sciences, generated discussions relevant to monotheistic religions. The exact sciences did not contain material at which religious authorities could take offense. In the West, it was always certain aspects of natural philosophy that prompted some religious authorities to react with fear and suspicion. The offensive parts, as we might expect, were those that conflicted with, or subverted, scripture.

I hope that I have made it clear that when I speak of the history of the relations between science and religion, I am almost always speaking about natural philosophy and theology. But who were the practitioners of natural philosophy and science in the ancient and medieval periods? What were they called? The only name that we recognize today is that of physician or doctor. Since the days of Hippocrates, all the way to the end of the Middle Ages, there have been recognized doctors and physicians. But the familiarity ends there. Because there was never a recognized class of scientists doing scientific research, there is no name for this group. To understand why this is so, we must understand how science and natural philosophy were done, and by whom. Doing science was quite different from doing natural philosophy.

What science there was in the period from the rise of Christianity to the end of the Middle Ages was carried out by individuals usually working alone and perhaps consulting with someone they might know who had similar interests. They would, of course, have read some of the literature relevant to the scientific topic or theme they were pursuing. Occasionally royal courts supported a person who had a reputation in optics, astronomy, or medicine, or in natural philosophy. Among the most famous of medieval royal patrons are Emperor Frederick II of the Holy Roman Empire (A.D. 1215–1250) and King Charles V of France (A.D. 1364–1380). Indeed, physicians could always be found at royal courts. Apart from medicine, science was a peripheral activity. Those who engaged in such activities usually supported themselves, probably in some professional capacity. They might have been teachers, or because they could read and write, they could have performed clerical services for a king or nobleman, or municipal government. Indeed, they might also have been clergymen, as Roger Bacon was. Those who did science were not called scientists; those who did geometrical optics were sometimes called perspectivists, largely because the discipline came to be called by the Latin term *perspectiva*. More often, they were not referred to by any disciplinary name.

Science in the late ancient and medieval periods was, however, radically different from modern science. Although some interesting experiments were carried out, they were relatively rare occurrences and certainly did not constitute a recognized aspect of scientific activity. Few claims were tested objectively. The experimental method did not yet exist. The mathematical sciences, however, were presented with the same kind of rigor as a modern treatise in mathematical physics. Treatises in geometrical optics and in statics were rigorously mathematical and based upon some empirical evidence. A thirteenth-century treatise on statics by Jordanus de Nemore, titled *On the Theory of Weights*, is a model of scientific procedure and mathematical rigor. It was from medieval versions of such works that the disciplines of optics and statics were greatly advanced in the sixteenth and seventeenth centuries. Medieval treatises in optics and statics form a legitimate part of any history of mathematical physics.

Religion and the Mathematical Sciences

The interrelationship between the exact mathematical sciences and religion and faith is decidedly one way: it was the exact sciences that could exert influence on theology and religion, but there was virtually no feasible way that religion could influence the content of the mathematical sciences. In statics, for example, Jordanus de Nemore's *On the Theory of Weights* is utterly devoid of religious content, or anything that might be construed as pertinent to religion. The same may be said of astronomical treatises, as, for example, Campanus of Novara's *Theory of the Planets* (*Theorica planetarum*), a thirteenth-century treatise that sought to present the great Ptolemaic astronomical system to the Latin Middle Ages. The book is a highly technical, mathematical account of Ptolemy's astronomy, which enabled many medieval scholars to learn about the quantitative motions of the planets (Campanus of Novara 1971). But it afforded no plausible opportunities to relate religion to astronomy, or vice versa. In optics, Roger Bacon wrote treatises that were either wholly on geometrical optics or in which geometrical optics played some role. And yet there is nothing about theology or religion in two of his major treatises, *On the Multiplication of Species* and *On Burning Mirrors* (Lindberg 1983). This is especially significant because Bacon firmly believed that the exact mathematical sciences were essential for theology and religion and that one could not be a knowledgeable theologian without studying these subjects. He was especially enthusiastic about geometrical optics because he believed the discipline was capable of illustrating spiritual truths. He was convinced, for example, that an incident ray is a certain kind of ray just as refracted rays and reflected rays of geometrical optics were useful in explicating and interpreting the spiritual infusion of grace (Bacon 1928, 1:238–239). But none of this played a role in his relevant optical treatises. Optics did, however, play a role in theology. Optical questions were discussed in basic theological treatises, in which theologians invoked the role of light in the creation and pursued a variety of themes about vision (see Lindberg 1976, 139–142). But these were instances of the application of geometrical optics, and to a greater extent natural philosophy, to theology. But optical discussions in theology had no effect or influence on geometrical optics during the late Middle Ages.

In contrast to the mathematical sciences, natural philosophy was directly affected by theology and religious doctrines, no doubt because it embraced motion and change in physical bodies throughout the universe. The central core of medieval natural philosophy consisted of a number of works by Aristotle that ranged over problems in physics, chemistry, meteorology, geology, cosmology, and biology. The numerous problems of late medieval natural philosophy were drawn from, or were based on, these Aristotelian works. Although natural philosophers wrote treatises on a wide variety of themes, the most widespread format, which was used in the universities, involved sequences of questions on one of Aristotle's treatises in natural philosophy.

THE PROPAGATION OF SCIENCE

The dissemination of science and natural philosophy in the ancient and medieval periods was a difficult and uncertain process. All treatises in late antiquity and the Middle Ages were written in ink on paper made from papyrus or rags. This remained so until the invention of movable type by Johann Gutenberg in the mid-fifteenth century. It is enormously difficult for modern readers who have never examined and studied medieval manuscripts to comprehend the monumental difficulties that were normal and routine. It will be useful to describe those difficulties.

During the Middle Ages in Western Europe, a copy of a treatise was usually made by a copyist, or scribe (see Figures 1.3 and 1.4). In the scribal culture of pre-printing days, scribes laboriously copied a work or treatise from a version they owned or to which they had access. Not only was this a slow process—despite a system of abbreviations that facilitated the process—but it was also virtually inevitable that the new copies would diverge from the original. No two copies would be identical. A scribe whose task it was to make five or ten copies of the same treatise would inevitably produce five or ten copies that differed in various ways from the original and from each other.

Over the years and centuries, popular works were copied many times. Modern scholars who wish to print editions of such works assemble as many of the manuscript copies of the work that they can find in the libraries of Europe and in private collections. When this is done, they inevitably discover that the copies differ from each other in many small and large ways. Words, and almost as often, whole sentences, are omitted from some copies. The words that are omitted from one manuscript copy will usually differ from the words omitted in another copy. The same may be said for whole sentences; sometimes entire paragraphs or pages are omitted and even whole sections of some works. The best results were probably obtained when numerous scribes were gathered together in one room and someone dictated the text to them. Greater uniformity and fewer errors were the probable results of this procedure. Such a method was employed largely at universities and colleges, where there was a ready market for the texts that were selected for dictation. Of hand-copied treatises and documents, we can confidently assume that a reader of a manuscript copy of a work in Paris knew that he was reading a variant version of the

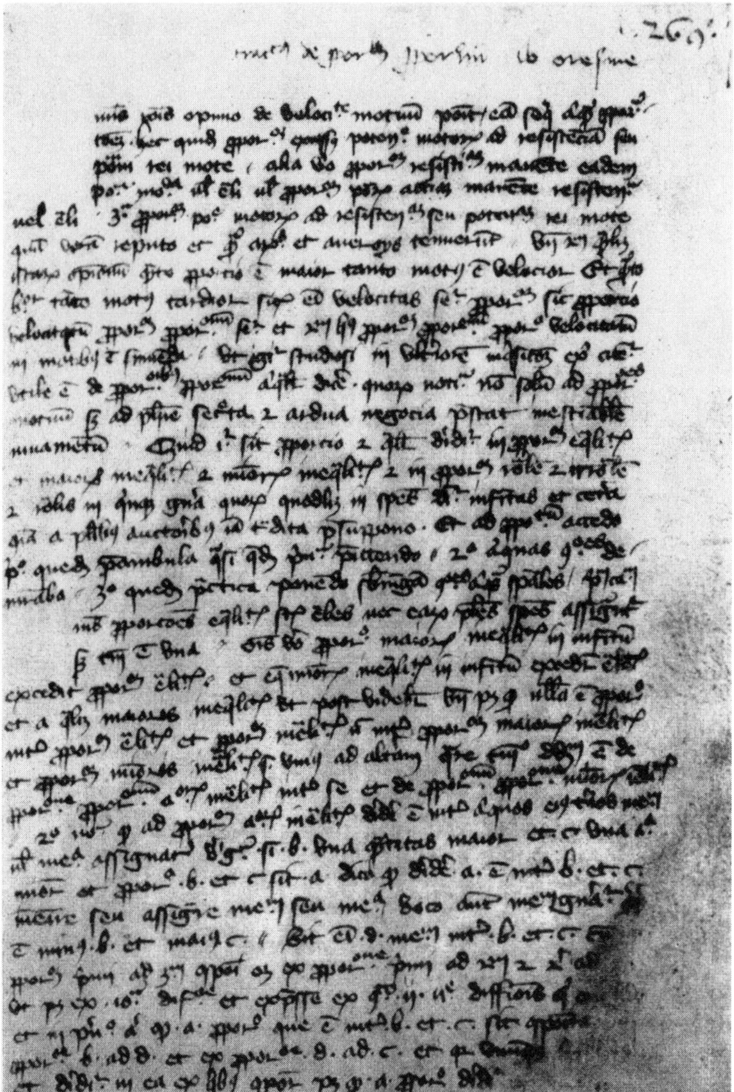

Figure 1.3. The opening page of Nicole Oresme's Latin treatise, *The Commensurability or Incommensurability of the Celestial Motions*. (From Paris, Bibliothèque Nationale, fonds latin, 7281, fol. 259r. The manuscript was copied sometime in the fifteenth century, probably after 1420.)

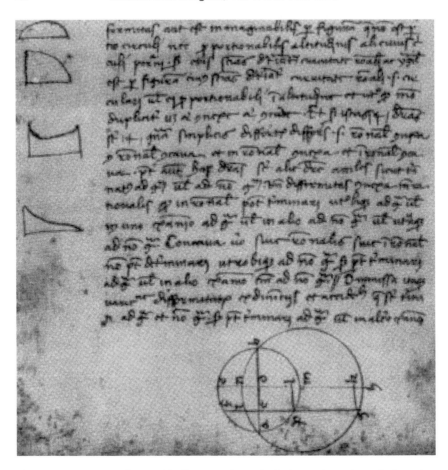

Figure 1.4.   A page of text, with diagrams, from Nicole Oresme's mathematical treatise, *On the Configurations of Qualities and Motions.* (From Florence, Biblioteca Nazionale Centrale, Fondo Coventi Soppressi J.IX.26, fol. 15v. The manuscript was copied around 1500.)

same treatise that another scholar was reading elsewhere in Paris, or perhaps in Oxford, Rome, Heidelberg, or Amsterdam. The differences between those texts might be large or relatively small. But one could be confident that they differed.

In scientific and mathematical texts, the dangers of scribal errors were much greater than in straightforward verbal texts. Diagrams and tables were essential parts of astronomical and mathematical texts. In copies of Euclid's *Elements,* for example, copies made and used by students and scholars in different European universities varied significantly. Certain diagrams might be omitted in some copies and included in others. Even diagrams that were included might be rendered unusable, or misleading, because the scribe mislabeled them. Moreover, there was the enormous problem of legibility. In many manuscript copies, the handwriting of the scribe is difficult to read or largely illegible. Under such circumstances, it is virtually self-evident that reliance on hand-copied manuscripts made doing science in the Middle Ages an extremely difficult task. It is rather remarkable that the study of science, natural philosophy, theology, and literature was as coherent and intelligible as it was. Scholars and students were somehow able to derive the essential features of the treatises they studied, read, taught, and wrote about. Although efforts were made to improve the quality of manuscript copies, it was virtually impossible to produce copies that were substantially identical. All implicitly recognized that the text of a given treatise would inevitably vary as more copies were made and more and more errors were introduced.

The advent of printing from movable type changed all this. One could now be confident that those reading scientific texts in Paris, Rome, and London were reading identical copies if those copies were printed by the same publisher in the same print shop. Printing from movable metal type made possible the dissemination of reliable knowledge on a massive scale in ways that were previously impossible. Books on science and medicine were now rapidly made available in identical copies throughout Europe. The plates, diagrams, and tables in a given book were the same in all copies. The advantages of printed books over hand-copied books cannot be overestimated. The printed book transformed learning in Europe not only because it introduced uniform standards, but also because it greatly increased the speed by which learning was disseminated. The invention of printing from movable type may have been the most important contribution to the advance of civilization made in the second millennium. The transition from hand-copied documents to printed documents was far more revolutionary than the transition from the typewriter to the computer.

Despite the difficulties of composing and disseminating treatises on science and natural philosophy during the late ancient and medieval periods, scholars have been able to construct a reliable account from the range of documents relevant to the theme of this book. To indicate the themes that will be pursued in this volume, I now present a thumbnail sketch of chapters two to eight.

BRIEF DESCRIPTIONS OF CHAPTERS 2–8

Chapter 2: Aristotle and the Beginnings of Two Thousand Years of Natural Philosophy

The relations between science and religion in the late ancient and medieval periods were intimately bound up with the works of Aristotle (384–322 B.C.), the great Greek philosopher and scientist of the fourth century B.C., to whom were attributed treatises on such diverse subjects as rhetoric, logic, poetry, politics, zoology, scientific method, and natural philosophy. Of these disciplines, natural philosophy, or physics, as it was frequently called, is the most significant for the relations between science and religion. As we have already seen, the exact sciences, such as mathematics, astronomy, optics, and mechanics, contained little that was offensive to religious authorities and theologians. But natural philosophy—Aristotle's natural philosophy—contained much about the physical cosmos that directly conflicted with Christian ideas about the world and its operation. To appreciate and understand this, it is essential to know who Aristotle was, what he wrote, and why he was the most important figure in the history of science and natural philosophy until the seventeenth century.

Chapter 3: Science and Natural Philosophy in the Roman Empire

I shall begin with a non-technical summary of Greek science as a whole from around 400 B.C. to the establishment of the Roman Empire. The status of science and natural philosophy in the period of the Roman Empire was the science and natural philosophy that Christianity encountered in the first centuries of its existence. Because Romans contributed little to science, Christians were surrounded essentially by Greek science and natural philosophy, both of which continued to be developed and advanced in this period.

Chapter 4: The First Six Centuries of Christianity: Christian Attitudes toward Greek Philosophy and Science

Christianity was born into an intellectual world that was dominated by secular Greek thought. Christians had a scripture that they began to explicate and interpret. Was that sufficient? What should their attitude be toward Greek science and natural philosophy, which gave a different interpretation of the world than did the Bible? Was it important for them to know about a view of the world that was essentially secular? Could such knowledge be of any use? The church fathers and other Christians debated these issues, and from those debates emerged the important concept of science and natural philosophy as "handmaidens of theology." That is, pagan, secular (Greek) science was to be studied and used only to the extent that it served the purposes of religion; it was not to be studied for its own sake. What is of great significance, however, is the fact that secular science could serve an important function for Christians: It would provide the tools to analyze God's creation.

Christians also found themselves using natural philosophy to explicate the creation account in Genesis. Such works were known as hexameral treatises, which literally means "the six days." An account of some of these important treatises will be included in this chapter.

Chapter 5: The Emergence of a New Europe after the Barbarian Invasions

After briefly describing conditions in Western Europe from approximately 600 to 1000, I shall depict the new intellectual circumstances that developed in the eleventh and twelfth centuries. This will include a description of the new cathedral schools and their curricula, which will be followed by a summary account of the new attitude that emerged toward traditional Church authority in the treatment of natural phenomena. An important feature of this new approach concerns dramatic developments in the approach to theology, developments that set the stage for theological commentaries over the next five centuries and which would involve natural philosophy.

Chapter 6: The Medieval Universities and the Impact of Aristotle's Natural Philosophy

I shall first briefly describe the influx of new knowledge into Western Europe by way of translations from Greek and Arabic into Latin. This will be followed by a general account of the structure and curriculum of the medieval university, primarily focused on the Universities of Paris and Oxford. For the remainder of the chapter, I shall focus on the negative reaction of some church authorities to Aristotle's natural philosophy, a resistance that culminated in the Condemnation of 1277 in Paris.

Chapter 7: The Interrelations between Natural Philosophy and Theology in the Fourteenth and Fifteenth Centuries

To describe the interrelations between natural philosophy and theology, I shall first explain how theology influenced natural philosophy, showing the degree to which theological and religious ideas penetrated natural philosophy, and then show the converse—the extent to which natural philosophy influenced theology.

Chapter 8: Relations between Science and Religion in the Byzantine Empire, the World of Islam, and the Latin West

I conclude the volume with a comparison of the way science and religion interacted in the Byzantine Empire, the Islamic world, and the Latin West. In this comparison, readers will see how the broad cultural and governmental structures of a civilization affect the way it shapes the relations between science and religion. A vital element in the long-term relationship between science and religion is the separation of church and state. Christians wanted the state to leave them alone to worship as they pleased. Hence they advocated a separation of church and state, which became an inherent feature of the relations between the two entities long after Christianity became the state religion. This stood in sharp contrast to the Byzantine Greek Orthodox Church and to Islamic Society. Other characteristics will also be explored, especially the role science and religion played in education.

Chapter 2

❖

Aristotle and the Beginnings of Two Thousand Years of Natural Philosophy

LIFE

Aristotle was born in 384 B.C., in the town of Stagira, which lay in Macedonia in northern Greece. His father, Nicomachus, was a physician in the service of King Amyntas of Macedon; his mother, Phaestis, was a woman of independent wealth (Barnes 1995, 3; Sarton 1952, 470–473). In 367, as a lad of seventeen years of age, Aristotle moved to Athens to study with Plato in the Academy, where he remained for twenty years until the death of Plato in 347. During those twenty years, it is plausible to assume that Aristotle heard, and participated in, important philosophical discussions involving some of the greatest minds of the time. The themes that were debated must surely have ranged across issues that were dear to Plato, such as metaphysics, ethics, logic, politics, and epistemology. And although physics and cosmology were not themes to which Plato devoted much time and effort, Aristotle would likely have had occasion to engage in discussions about those subjects.

With the death of Plato and the emergence in Athens of anti-Macedonian sentiment, Aristotle, who never became an Athenian citizen, departed Athens and traveled to the coast of Asia Minor. There he lived first in Assos, where he married Pythias, the niece of the tyrant of Assos, Hermias (Sarton 1952, 473). He moved next to Mytilene, on the island of Lesbos, where he met Theophrastus, who became an important friend and future colleague. During his approximately four years in this region, it is likely that Aristotle studied marine biology and later used what he learned to write his biological treatises.

In 343, after a brief return from Mytilene to his home in Stagira, Aristotle received an invitation from Philip II, King of Macedon, to tutor his son, the future Alexander the Great. "Thus," declares Jonathan Barnes, "began the association between the most powerful mind of the age and the most powerful man" (Barnes 1995, 5). Although one could all too easily imagine how these two great figures might have mutually interacted and shaped each other's thoughts and actions, it is well to keep in mind Barnes' admonition that "what Aristotle said to Alexander the Great, and Alexander to him, we do not know" (Barnes 1995, 5).

After a long absence from Athens, Aristotle returned in 335 and established the Lyceum, a rival school to Plato's Academy. It was here where most of Aristotle's extant works were written. On the occasion of the death of Alexander the Great in 323, Aristotle once again fell victim to anti-Macedonian rage. In 322, he fled to Chalcis on the island of Euboea, where he died the same year at the age of sixty-two.

To breathe some life into this skeletal account, here is a vivid description of Aristotle the man:

Of Aristotle's character and personality little is known. He came from a rich family. He was a bit of a dandy, wearing rings on his fingers and cutting his hair fashionably short. He suffered from poor digestion, and is said to have been spindle-shanked. He was a good speaker, lucid in his lectures, persuasive in conversation and he had a mordant wit. His enemies, who were numerous, made him out to be arrogant and overbearing. His will, which has survived, is a generous and thoughtful document. His philosophical writings are largely impersonal; but they suggest that he prized both friendship and self-sufficiency, and that, while conscious of his place in an honourable tradition, he was properly proud of his own attainments. As a man, he was, I suspect, admirable rather than amiable. (Barnes 1982, 1)

WORKS

Aristotle left behind a substantial legacy of written works. In *The Complete Works of Aristotle*, forty-six works are included, of which sixteen are deemed spurious by modern scholars (Aristotle 1984). Drawing upon different ancient sources in which both lost and extant works are mentioned, it appears that Aristotle may have written as many as 200 works (see Aristotle 1984, 2:2386–2388). For obvious reasons, however, scholars have focused their attention on the thirty works regarded as genuinely by Aristotle. Let us see how these works are regarded.

At the outset, we confront a puzzle about the true nature of the Aristotelian texts. Upon Aristotle's death, his library, which included a large number of treatises ostensibly written by him, passed to his friend Theophrastus, who became the new head of the Lyceum, and who subsequently passed the library to his nephew, Neleus, who took it to Scepsis, a city in Asia Minor. There, Neleus buried the collection in a cave. For two centuries, the manuscripts lay rotting in that cave and were then rediscovered and taken to Athens, then to Rome, where they came into the possession of Andronicus of Rome (first century B.C.). Andronicus, who was an Aristotelian philosopher, edited the manuscripts around 70 B.C. and in so doing prepared the basic edition that is still in use today. Jonathan Barnes has asked, "What did Andronicus do? How did his edition—how does our edition—differ from what Aristotle actually wrote? The answer roughly put, is probably this: Andronicus himself composed the works which we now read" (Barnes 1995, 10–11; also Montgomery 2000, 7–9, for further details).

How are we to understand this startling statement? Andronicus did not actually write the works, but by editing them, he gave them the form they have today. The works in their present form were thus not written by Aristotle, or even by his pupils and colleagues in the Lyceum. Who, then, did compose the original works that Andronicus edited? What did Aristotle actually write? These questions have been hotly debated since the nineteenth century. We shall not attempt to answer them here. It has been suggested that Aristotle's works in the form we have them were probably lecture notes or drafts of his lectures. Over the years, and for a variety of reasons, Aristotle undoubtedly made changes in his notes, which may account for the unpolished and uneven style of his texts. Later members of the Lyceum may have edited some of his treatises, and Andronicus of Rhodes certainly did.

There is also the problem of the chronology of Aristotle's works. What was the order of his works? Can this be determined? For numerous reasons, it is not likely that we shall ever be able to ascertain the chronology of his works with any reasonable degree of confidence. Questions about Aristotle's texts—authorship, chronology, and authenticity—emerged in the late nineteenth century and have attracted the attention of scholars ever since. But in a study about the history of natural philosophy, in which Aristotle is the paramount figure, those problems are of little consequence. During the lifetime of Aristotelianism as a viable philosophy, and even beyond into the nineteenth century, such considerations were never raised. It did not occur to medieval scholars, for example, to question the authorship of such texts. Even if they had known that Andronicus of Rhodes edited the works of Aristotle, they would not have found that any reason to question the veracity of Aristotle's authorship. Moreover, they were never concerned about the consistency of this or that particular treatise; nor did they inquire whether someone—Andronicus, for example—might have rearranged and interpolated passages. No one suggested that Aristotle's treatises were really the collective product of many minds at the Lyceum. Such problems and questions never arose. Until the late nineteenth century, the great philosopher Aristotle, regarded as the dominant intellect of antiquity and the Middle Ages, was assumed, without question, to have written the texts attributed to him.

In fact, no one to my knowledge raised questions about the chronology of Aristotle's works. In the modern age, it has been rightly assumed that Aristotle's ideas must have evolved and that this would be somehow reflected in his works. If statement *A* in one work conflicts with statement *B* in another treatise, modern scholars assume that Aristotle changed his mind. Thus, if one can determine that statement *B* was made subsequent to statement *A*, it would follow that the section containing *B* was written after the section containing *A*, and, therefore, perhaps the work including section *B* was written after the work that includes section *A*.

Questions about the chronology of Aristotle's works did not really arise until the late nineteenth century. His works were approached as if they had been composed in a timeless manner. Inconsistencies of the kind described in the preceding paragraph would have been resolved by somehow reconciling the two conflicting statements, or by ignoring them. And above all, despite the legitimate concerns of modern scholars about the nature of Aristotle's authorship, or whether he should even be regarded as the author of the many treatises attributed to him, philosophers and scientists who used them in late Greek antiquity, in Islam, and in Western Europe from the late Middle Ages to the end of the nineteenth century, assumed, without reservation, that the philosopher Aristotle was the undoubted author of them all.

ACHIEVEMENTS

Aristotle is probably the most significant figure in the history of Western thought up to the end of the sixteenth century. The range of topics he treated in his extant writings is extraordinary, and the wisdom and insight he reveals is rather amazing for someone who lived in the fourth century B.C. We can best appreciate Aristotle's contributions when we view them against the background of his role in Western thought. As G.E.R. Lloyd explains: "To attempt to cover the history of Aristotle's influence on subsequent thought in full would be not far short of undertaking to write the history of European philosophy and science, at least down to the sixteenth century" (Lloyd 1968, 306). Aristotle's contributions are nothing less than monumental. Lloyd rightly declares that "the idea of carrying out systematic research is one that we in the West owe as much to Aristotle and to the Lyceum as to any other single man or institution" (Lloyd 1968, 287). Aristotle's research programs are exhibited in his biological and political works. Indeed, he is the founder of biology as a discipline. Not only did he do pioneering work in biological classification, but his description of the habits and behavior of certain species still elicit the admiration, and even awe, of modern zoologists.

In the *History of Animals*, we see Aristotle at his best as an observer and recorder of animal behavior. Among his numerous especially noteworthy descriptions are those of the torpedo fish, or stingray, the breeding habits of bees, as well as embryological data about the chick, the placental shark, and cephalopods (Aristotle 1984, 966; 970–976; 883–884; 889; 855–856; see also Sarton 1952, 537–545). Aristotle examined chick embryos, for example, by breaking open one egg every day and observing the progress of the embryo. On the third day, he observed the beginning of an embryo and noted that "the heart appears, like a speck of blood in the white of the egg"; on the tenth day, "the chick and all its parts are distinctly visible" (Aristotle 1984, 883). Although Aristotle knew that most fish produce their young by laying eggs, he recognized an important exception in the placental dog fish, a member of the selachian group, which brings forth its young alive. It does this by laying eggs in the womb, whereupon the eggs become attached to the womb by a navel string (Cohen and Drabkin 1948, 420). Most were skeptical of Aristotle's claim until 1842, when Johannes Müller, the great German biologist, showed that Aristotle's observations were correct (Sarton 1952, 541–542). Aristotle described the internal parts of 110 animals, of which he may have dissected as many as 49, perhaps even an elephant (Lones 1912, 105–106). At the very least, Aristotle dissected the eye of a chick, the eye of a mole, and the stomachs of ruminants, which he accurately describes in his *History of Animals* (Aristotle 1984, 883; 844; 806; see also Cohen and Drabkin 1948, 412–413).

Aristotle's biological works were still influential and current in the nineteenth century, as two tributes to him reveal. George Henry Lewes (A.D. 1817–1878), a naturalist who carefully studied Aristotle's biological treatises, was particularly impressed by Aristotle's *On the Generation of Animals*, of which he said the following:

It is an extraordinary production. No ancient, and few modern works, equal it in comprehensiveness of detail and profound speculative insight. We there find some of the obscurest problems of Biology treated with a mastery which, when we consider the condition of science at that day, is truly astounding. That there are many errors, many deficiencies, and not a little carelessness in the admission of facts, may be readily imagined; nevertheless, at times the work is frequently on a level with, and occasionally even rises above, the speculations of many advanced embryologists. (Lewes 1864, 325; cited in Sarton 1952, 540)

The second tribute to Aristotle was from none other than Charles Darwin. Upon receipt of a copy of Dr. William Ogle's translation of Aristotle's *Parts of Animals*, Darwin sent a letter of thanks to Ogle on February 22, 1882: "You must let me thank you for the pleasure which the introduction to the Aristotle book has given me," began Darwin, who then declared (Sarton 1952, 545):

I have rarely read anything which has interested me more, though I have not read as yet more than a quarter of the book proper.

From quotations which I had seen, I had a high notion of Aristotle's merits, but I had not the most remote notion what a wonderful man he was. Linnaeus and Cuvier have been my two gods, though in very different ways, but they were mere schoolboys to old Aristotle. How very curious, also, his ignorance on some points, as on muscles as the means of movement. I am glad that you have explained in so probable a manner some of the grossest mistakes attributed to him. I never realized, before reading your book, to what an enormous summation of labour we owe even our common knowledge. I wish old Aristotle could know what a Defender of the Faith he had found in you. Believe me, my dear Dr. Ogle,

Yours very sincerely,
Ch. Darwin

If Aristotle is the founder of biology, he is also the universally acknowledged inventor of formal, syllogistic logic. His logical treatises are known as the Organon—tool or instrument—and consist of six works: the *Categories*, which treats of terms that signify things; *On Interpretation*, which is concerned with propositions; *Topics*, in eight books, which is concerned with non-demonstrative reasoning and shows how to conduct effective arguments; *Sophistical Refutations*, which is a collection of fallacies. The final two, the *Prior Analytics* and the *Posterior Analytics*, are the most important. The *Prior Analytics* contains Aristotle's greatest contribution to the history of logic, namely the theory of the syllogism, which is the theory of deductive inference and usually consists of two premises and a conclusion. Not only was Aristotle the first to present a formal analysis of the syllogism, but he also "invented the use of schematic letters" (Barnes 1982, 30). As Jonathan Barnes explains, "Logicians are now so familiar with his invention, and employ it so unthinkingly, that they may forget how crucial a device it was: without the use of such letters logic cannot become a general science of argument. The *Prior Analytics* makes constant use of schematic letters" as is evident early on when Aristotle presents the classic case of the syllogism: "If A is predicated of every B, and B of every C, A must be predicated of every C" (Aristotle 1984, 41). Finally, there is the *Posterior Analytics*, wherein Aristotle presented his theory of scientific demonstration or scientific method and used the mathematical sciences as his primary model.

In addition to his work in biology and logic, Aristotle constructed a system of the cosmos that endured for more than 2,000 years, in three different civilizations and cultures. His discussions of motion in the *Physics* set the stage for subsequent controversies that resulted in medieval advances and ultimately led to the works of Galileo and Newton. In his brief *Poetics*, Aristotle established the categories of drama that are still accepted, namely "Tragedy, Comedy, Epic, and Lyric" (Oesterle 1967, 455). Within the category of Tragedy he distinguished six elements: "plot, character, thought, diction, song, and spectacle" (Oesterle 1967, 456). As one author put it, "after twenty-two centuries it remains, the most stimulating and helpful of all analytical works dealing with poetry" (Cooper 1956, 3). His ideas about politics and ethics formed the basis of discussions in those areas until early modern times.

It is difficult to overstate Aristotle's significance for Western thought. His contributions can be summarized in many ways. A brief account by A. E. Taylor will suffice:

It has not been the lot of philosophers, as it is of great poets that their names should become household words. . . . Yet there are a few philosophers whose influence on thought and language has been so extensive that no one who reads can be ignorant of their names, and that every man who speaks the language of educated Europeans is constantly using their vocabulary. Among this few Aristotle holds not the lowest place. We have all heard of him, as we have all heard of Homer. He has left his impress so firmly on theology that many of the formulae of the Churches are unintelligible without acquaintance with his conception of the universe. If we are interested in the growth of modern science we shall readily discover for ourselves that some knowledge of Aristotelianism is necessary for the understanding of Bacon and Galileo, and the other great anti-Aristotelians who created the "modern scientific" view of Nature. If we turn to the imaginative literature of the modern languages, Dante is a sealed book, and many a passage of Chaucer and Shakespeare and Milton is half unmeaning to us unless we are at home in the outlines of Aristotle's philosophy. And if we turn to ordinary language, we find that many of the familiar turns of modern speech cannot be fully understood without a knowledge of the doctrines they were first forged to express. An Englishman who speaks of the "golden mean" or of "liberal education," or contrasts the "matter" of a work of literature with its "form," or the "essential" features of a situation or a scheme of policy with its "accidents," or "theory" with "practice," is using words which derive their significance from the part they play in the vocabulary of Aristotle. (Taylor 1955, 5–6)

Of the numerous themes that one might investigate in the thought of Aristotle, we shall focus our attention on his natural philosophy, which served as the dominant interpretation of nature for approximately 2,000 years, encompassing at least three civilizations, using three distinct languages: Greek, Arabic, and Latin.

ARISTOTLE'S COSMOS AND NATURAL PHILOSOPHY

Aristotle's cosmos was, of course, a product of his natural philosophy. By a variety of means and in a number of treatises, Aristotle identified and described the basic components of the physical world. What he fashioned was destined to serve as the basic conception of the universe for almost two millennia.

For Aristotle, the cosmos was a gigantic spherical plenum that had neither a beginning nor would ever have an end. Everything in existence lies within that sphere; nothing exists, or can possibly exist, outside of it: neither matter, nor empty space, nor time, nor place. Aristotle regarded it as nonsensical to inquire about extracosmic existence, consequently rejecting the possibility that other worlds might exist beyond our own. Within the cosmos, Aristotle distinguished two major divisions: the celestial region and the terrestrial. The dividing line between the two regions was the concave surface of the lunar sphere. That surface divided two totally dissimilar regions (see Figure 2.1).

The terrestrial region, which lay below the concave lunar surface was a region of constant change and transformation. It consisted of four elements: earth, water, air, and fire, arranged in this order from the center of the world to the moon's concave surface. All bodies were compounded of combinations of two or more elements. In the terrestrial region, bodies were always coming into being as differing compounds of the four elements, and bodies were always passing away because their elements eventually dissociated to combine with other elements and form new compound bodies. At the center of the universe was the earth, surrounded in many of its parts by water and then air and fire. If the motions of the elements were suddenly to cease, the four elements would sort themselves out into four concentric regions, from heaviest to lightest, namely from earth to water to air to fire. But this could not happen because it is the nature of all elements to move and thereby to associate and dissociate with other elements. In the upper atmosphere of the terrestrial region, just below the concave surface of the moon, Aristotle assumed that comets, shooting stars, and other similar phenomena occurred. He inferred their existence in this region because they were changeable phenomena, and therefore could not occur in the unchanging celestial region.

If change and transformation are the characteristic features of the terrestrial area, minimal change is the hallmark of the celestial region,

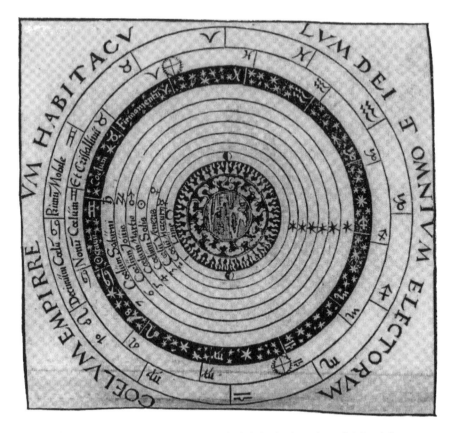

Figure 2.1. The basic, simplified medieval Christianized version of Aristotle's cosmos, depicting the terrestrial region in the center and the planetary spheres from the lunar orb to the "first movable heaven" (*primum mobile*). Enclosing the world is the immobile empyrean heaven, the "dwelling place of God and all the elect." (Peter Apian, *Cosmographicus liber*, 1524, col. 6. Courtesy Lilly Library, Indiana University, Bloomington.)

within which lie the planets and stars. The lack of change is attributable to a celestial ether, which Aristotle regarded as a fifth element that fills the celestial region, leaving no empty spaces. The ether is an incorruptible, eternal substance that suffers no change, except change of place. It did not come into being and will never pass away. Because the planets and stars are composed of the celestial ether, they also undergo no change, except change of place, which we can readily observe. Because Aristotle viewed a small degree of change as superior to a greater degree of change, he regarded the celestial region as nobler than, and vastly superior to, the terrestrial region, where incessant and unremitting change was the most characteristic feature. Because it is nobler and superior to the terrestrial region, Aristotle thought it appropriate that the celestial region should influence terrestrial changes. Future astrologers found this a welcome support to justify their prognostications.

It was Aristotle's understanding of natural philosophy that enabled him to determine the physical nature of the cosmos and to spell out its properties and behavior in his numerous treatises. To grasp the role that natural philosophy played in Aristotle's scheme of things, it is essential to understand the emphasis he placed on human reason, or, intellect, which is the same thing. In the *Nicomachean Ethics* (10.7.1178a.5–8), Aristotle declares that "that which is proper to each thing is by nature best and most pleasant for each thing; for man, therefore, the life according to intellect is best and pleasantest, since intellect more than anything else *is* man" (Aristotle 1984, 1862). Aristotle frequently emphasized reasoned discourse and accorded it the highest place. Although he did not assume a creation for the world, he did believe in a God, but a rather strange God, one who serves as a final cause for an eternal world, without beginning or end. Indeed, Aristotle's God has no knowledge of our world's existence, but is wholly absorbed in thinking about himself, since he alone is worthy of serving as his own object of thought. Even if the world were not the object of God's thoughts, Aristotle regarded it as a rationally structured physical sphere that contained all that exists, with nothing lying beyond.

Aristotle thought it important to classify different kinds of knowledge and actions in appropriate categories. Where, then, did he locate natural philosophy, or natural science, within the all-inclusive domain of knowledge? In his *Metaphysics*, Aristotle distinguished three broad categories of knowledge that he regarded as scientific: the productive sciences, the practical sciences, and the theoretical sciences. The productive sciences embraced all knowledge concerned with the making of useful objects, whereas the practical sciences were directed toward human conduct. Everything else fell under the jurisdiction of the theoretical sciences, which Aristotle divided into three parts (Aristotle 1984, 1619–1620). If we take them in order of priority, they are: (1) metaphysics, or theology, which considers things that are unchangeable and therefore distinct and separable from matter or body, such as God and spiritual substances; (2) mathematics, which also considers things that are unchangeable. Unlike objects in metaphysics, however, the objects of mathematics have no separate existence because they are abstractions from physical bodies; and (3) physics, often called natural science, or, as it came to be popularly designated, natural philosophy, which considers only those things that are changeable, exist separately, and also have within themselves an innate source of movement and rest. From Aristotle's standpoint, natural philosophy embraces both animate and inanimate bodies and is applicable to the whole physical world, that is, to both the terrestrial and celestial regions.

But how do we derive knowledge about nature? First it is essential to understand that for Aristotle, as he explains in the *Posterior Analytics*, sense perception is the foundation of human knowledge. He regards it as "impossible to get an induction without perception—for of particulars there is perception" (Aristotle 1984, 132). But as basic as he regarded sense perception, Aristotle denied that we could arrive at scientific knowledge by means of sense perception alone. To attain to scientific knowledge, universal propositions are essential. One arrives at universals from sense perceptions by means of induction (Aristotle 1984, 132). It is by means of universals, not direct perception, that we can generate demonstrations that produce scientific knowledge. As Aristotle explains, perception is of the individual or particular, and from particulars "it is impossible to perceive what is universal and holds in every case" (Aristotle 1984, 144). But demonstrations that yield scientific knowledge are based on universal propositions. That is why "if we were on the moon and saw the earth screening it we would not know the explanation of the eclipse. For we would perceive that it is eclipsed and not why at all; for there turned out to be no perception of the universal. Nevertheless, if, from considering this often happening, we hunted the universal, we would have a demonstration; for from several particulars the universal is clear" (Aristotle 1984, 144).

Although Aristotle placed great emphasis on demonstrations based on universal propositions arrived at inductively and ultimately from sense perceptions of individual things, he was equally interested in explaining the causes of natural phenomena, even phenomena not directly perceived. In the beginning of the seventh chapter of the first book of his *Meteorology*, Aristotle explains that "We consider a satisfactory explanation of phenomena inaccessible to observation to have been given when our account of them is free from impossibilities" (Aristotle 1984, 562). Thus, even if one cannot directly observe a phenomenon, Aristotle feels that an explanation is acceptable if it is compatible with what is possible. His explanations of the formation of comets and shooting stars exemplify this procedure.

In addition to general methodological considerations, Aristotle placed great emphasis on the role of causes in natural philosophy. He regarded all bodies as composites of matter and form, the former functioning as a passive principle, the latter as an active principle. How do these bodies change? Aristotle attributed all possible changes to four kinds of causes. The first is the material cause—the matter from which something is made, as the bronze is the matter of a bronze statue. The second is the formal cause—the essence or inner structure of a thing as expressed in its definition. To pursue the statue example, the sensible aspect of form is the shape the sculptor will give to the statue; the intelligible aspect of form in this case would be the essence of what it is to be a statue. The third type is the efficient cause, which is the agent or producer of the change or action, namely the sculptor. The fourth is the final cause, which is the end or purpose for which an action is done. In the present case, the final cause of the bronze statue is the sculptor's original intent, for it was what motivated the sculptor to make the statue. Aristotle sometimes reduced the four causes to two. The material cause always remained the material cause. The other three causes can be reduced to a single cause, a formal-final-efficient cause. Thus, if an artist has the formal and final causes of a statue in mind, they will serve as an efficient cause to prompt him or her to make the statue. Or to use an organic example, an acorn does not yet have the form of an oak tree, although it has the potentiality of becoming an oak tree. Thus, the acorn will try to realize the form of an oak tree. Its ultimate goal of becoming an oak tree also functions as a final cause. The efficient cause operates to enable the acorn to realize its final form as an oak tree.

Aristotle distinguished four kinds of changes that these causes could produce. The most fundamental was substantial change in which one form replaces another form in the underlying matter of the body, as when fire reduces a log to ashes. Qualitative change occurs when, for example, the color of a body changes, as when a leaf turns from green to brown. Quantitative change takes place when a body increases or diminishes its size, but retains its identity. Finally, there is change of place, when a body moves from one place to another.

Aristotle had other tools of analysis for comprehending nature. In the *Physics,* he contrasts things that exist by nature with those that do not. "By nature the animals and their parts exist, and the plants and the simple bodies (earth, fire, air, water)—for we say that these and the like exist by nature" (Aristotle 1984, 329). Each of these things "has within itself a principle of motion and of stationariness (in respect of place, or of growth and decrease, or by way of alteration). On the other hand, a bed and a coat and anything else of that sort, *qua* receiving these designations—i.e. in so far as they are products of art—have no innate impulse to change" (Aristotle 1984, 329). But products of art will undergo change if they are composed of things that do have an impulse to change.

Later, in the *Physics,* Aristotle characterizes nature as "a cause, a cause that operates for a purpose" (Aristotle 1984, 341) and then defines it as "a principle of motion and change" (Aristotle 1984, 342). Thus, nature operates by causes that produce motions and changes. An investigation of nature by means of physics, or natural philosophy, would involve a study and analysis of those causes and the motions and changes they produce. In the introductory paragraph to his *Meteorology*, Aristotle gives us a good sense of what we should understand of the study of nature by natural philosophy.

We have already discussed the first causes of nature, and all natural motion, also the stars ordered in the motion of the heavens, and the corporeal elements—enumerating and specifying them and showing how they change into one another—and becoming and perishing in general. There remains for consideration a part of this inquiry which all our predecessors called meteorology. It is concerned with events that are natural, though their order is less perfect than that of the first of the elements of bodies. They take place in the region nearest to the motion of the stars. Such are the milky way, and comets, and the movements of meteors. It studies also all the affections we may call common to air and water, and the kinds and parts of the earth and the affections of its parts. These throw light on the causes of winds and earthquakes and all the consequences of their motions. Of these things some puzzle us, while others admit of explanation in some degree. Further, the inquiry is concerned with the falling of thunderbolts and with whirlwinds and fire-winds, and further the recurrent affections produced in these same bodies by concretion. When the inquiry into these matters is concluded let us consider what account we can give, in accordance with the method we have followed, of animals and plants, both generally and in detail. When that has been done we may say that the whole of our original undertaking will have been carried out. (Aristotle 1984, 555)

From this it is apparent that for Aristotle, physics, or natural philosophy, embraces the motions of terrestrial and celestial bodies, the motions and transformations of the four elements in the terrestrial region, and the generations and corruptions of the compound bodies they continually produce. Natural philosophy also includes phenomena in the upper region of the atmosphere just below the moon, which was Aristotle's concern in the *Meteorology*. Finally, it includes the study of animals and plants, which Aristotle says he will subsequently present. These, and other topics on natural philosophy, appear in a collection of Aristotle's treatises that came to be known collectively as the "natural books" (*libri naturales*), which include *Physics, On the Heavens (De caelo), On the Soul (De anima), On Generation and Corruption (De generatione et corruptione), Meteorology,* and *The Short Physical Treatises (Parva naturalia),* which consists of a number of brief treatises titled: *Sense and Sensibilia; On Memory; On Sleep; On Dreams; On Divination in Sleep; On Length and Shortness of Life; On Youth, Old Age, Life and Death, and Respiration* (see Aristotle 1984).

THE SCOPE OF NATURAL PHILOSOPHY

Natural philosophy in its many manifestations was practiced long before Aristotle made his momentous contribution. It is apparent in Egyptian civilization and among Aristotle's predecessors, the pre-Socratic philosophers. But, as far as is known, no one in those places and times sought to define anything resembling what we might regard as natural philosophy. They simply wrote about a variety of topics, and it has fallen to modern historians to decide whether what they wrote ought to be categorized as natural philosophy. Since it was not excluded in ancient Egypt or in Greece in the sixth and fifth centuries B.C., it seems appropriate to include medicine within the domain of natural philosophy, and, perhaps, even magic as well, though the latter would more properly form part of natural philosophy in ancient Egypt than in the Greece of the pre-Socratics.

Aristotle's contributions to natural philosophy changed all this. Not only did he leave treatises on almost all aspects of natural philosophy, but he also realized the need to define natural philosophy and delineate its scope, as well as to determine the best methodology for applying it to nature. Aristotle was apparently the first to perform this service. His efforts were destined to have a lasting effect, enduring for nearly 2,000 years in three different major linguistic cultures and civilizations—Greek (Byzantine Empire), Arabic (Islamic civilization), and Latin (Western Europe).

How did Aristotle define and understand natural philosophy? We have already seen that by defining it and enumerating the range of subjects to which it applies (in the *Meteorology*), he restricted its scope. This is obvious by his division of the theoretical sciences into metaphysics, mathematics, and natural philosophy, or physics. Clearly, he thought of metaphysics and mathematics as distinct from natural philosophy. Their subject matter was with entities that did not suffer change, while the essence of natural philosophy was to treat wholly of bodies undergoing change and motion. But did Aristotle really mean all bodies subject to change and motion? If so, natural philosophy would embrace virtually every discipline that treats of some aspect of the physical world, every part and subdivision of which undergoes change and motion. Because medicine is concerned with changes in the human body, it is appropriate to infer that Aristotle included medicine as part of natural philosophy. But this seems unlikely. In the opening passage of *Meteorology* (cited earlier), Aristotle intended to mention, or allude to, all the subjects that formed part of his research program. We may infer this from his remark that when the study of animals and plants has been completed, "we may say that the whole of our original undertaking will have been carried out." Nowhere in that "original undertaking" is medicine mentioned, nor, as far as we know, did Aristotle ever write a treatise on medicine, although he often used examples from medicine and was the son of a physician.

In addition to the exclusion of medicine from natural philosophy, Aristotle also excluded the mathematical, or exact, sciences, which he characterized as "the more natural of the branches of mathematics, such as optics, harmonics, and astronomy" (Aristotle 1984, 331). Some lines earlier, Aristotle explained that when, for example, a mathematician treats of celestial bodies, he does not "treat of them as the limits of a natural body; nor does he consider the attributes indicated [that is, the shapes of celestial bodies] as the attributes of such bodies. That is why he separates them; for in thought they are separable from motion, and it makes no difference, nor does any falsity result, if they are separated" (Aristotle 1984, 331). As we saw, Aristotle regarded optics, astronomy, and harmonics as "the more natural of the branches of mathematics," and therefore seemingly belonging more to mathematics than to natural philosophy. These sciences are "the converse of geometry. While geometry investigates natural lines, but not *qua* natural, optics investigates mathematical lines, but *qua* natural, not *qua* mathematical" (Aristotle 1984, 331). For Aristotle, the exact mathematical sciences fell somewhere between natural philosophy and pure mathematics, perhaps closer to the latter than the former. But the exact sciences belong neither wholly to natural philosophy nor to mathematics, but are relevant to both. Because they were viewed as lying between the two disciplines, the exact sciences came to be known as *middle sciences* (*scientiae mediae*) during the Middle Ages.

Aristotle's Approach

If Aristotle furnished the content, scope, and methodology of natural philosophy, he also provided something else of almost equal importance: a positive attitude toward nature and a style of doing natural philosophy. If we use hindsight to categorize Aristotle, we would judge that he had an intellectual temperament that was forged from the combined qualities of a philosopher, scientist, and historian. The historian in Aristotle is manifested in the way he presented problems. In the first book of the *Metaphysics*, Aristotle became the first historian of philosophy when he set forth the opinions of his predecessors, those natural philosophers who are known as the pre-Socratic philosophers, including his teacher, Plato. The problem he was investigating concerned the substance of things: What was their elemental nature? What was the cause of things? "We have studied these causes sufficiently in our work on nature," Aristotle declares, "yet let us call to our aid those who have attacked the investigation of being and philosophized about reality before us. For obviously they too speak of certain principles and causes; to go over their views, then will be of profit to the present inquiry, for we shall either find another kind of cause, or be more convinced of the correctness of those which we now maintain" (Aristotle 1984, 1555). With this said, Aristotle launches on a discussion of his predecessors for the rest of the first book.

Aristotle again resorts to the opinions of his predecessors in his cosmological treatise, *On the Heavens*. After asking "whether the heaven is ungenerated or generated, indestructible or destructible," he declares:

Let us start with a review of theories of other thinkers; for the proofs of a theory are difficulties for the contrary theory. Besides, those who have first heard the pleas of our adversaries will be more likely to credit the assertions which we are going to make. We shall be less open to the charge of procuring judgement by default. To give a satisfactory decision as to the truth it is necessary to be rather an arbitrator than a party to the dispute. (Aristotle 1984, 463)

Aristotle was well aware that natural philosophy had a history, and he appealed to it in numerous places.

But the roles of historian, philosopher, and scientist are not easily distinguished in Aristotle's treatises. When Aristotle invokes history and his predecessors, it is not merely for historical reasons, but rather to set the stage for resolving important scientific and philosophical issues. Thus, in the first chapter of the third book of his *Metaphysics*, he once again cites other opinions:

These include both the other opinions that some have held on certain points, and any points besides these that happen to have been overlooked. For those who wish to get clear of difficulties it is advantageous to state the difficulties well; for the subsequent free play of thought implies the solution of the previous difficulties, and it is not possible to untie a knot which one does not know. But the difficulty of our thinking points to a knot in the object; for in so far as our thought is in difficulties, it is in like case with those who are tied up; for in either case it is impossible to go forward. Therefore one should have surveyed all the difficulties beforehand, both for the reasons we have stated and because people who inquire without first stating the difficulties are like those who do not know where they have to go; besides, a man does not otherwise know even whether he has found what he is looking for or not; for the end is not clear to such a man, while to him who has first discussed the difficulties it is clear. Further, he who has heard all the contending arguments, as if they were parties to a case, must be in a better position for judging. (Aristotle 1984, 1572–1573)

Aristotle then lays out various problems and difficulties that he will subsequently consider. He usually sought to identify all the problems relevant to any issue he attempted to resolve (Grant 2001, 91–94).

To see how Aristotle proceeded in his analysis of problems in natural philosophy, it will be helpful to present a few examples. In the fourth book of his *Physics*, Aristotle argued strongly against the possibility that a vacuum could exist. As he so often did, Aristotle thought it best to begin by presenting the conflicting viewpoints, declaring that "We must begin the inquiry by putting down the account given by those who say that it exists, then the account of those who say that it does not exist, and third the common opinions on these questions"

(Aristotle 1984, 362; Grant 1981, 5–8). After setting out the conflicting opinions, Aristotle, in a typical move, declared, "As a step towards settling which view is true, we must determine the meaning of the word." He then proceeds to present different ways that the term "void," or vacuum, has been defined and conceived (Aristotle 1984, 363–365). Finally, he offers a series of arguments to demonstrate the impossibility of the existence of void space (Aristotle 1984, 365–369).

We see the same concern for the meaning of crucial terms in his cosmological work, *On the Heavens*. In this treatise, Aristotle asks whether there is only one heaven, or world, and whether it is eternal. In order to answer such questions, Aristotle insists that "we must explain what we mean by 'heaven' and in how many ways we use the word, in order to make clearer the object of our inquiry" (Aristotle 1984, 462). He then distinguishes three different usages of the term "heaven," or world. In one sense, world is taken as equivalent to the outermost circumference of the whole world; in another, heaven is conceived to embrace the whole celestial region, including the moon, sun, and the other celestial bodies; and, finally, heaven is taken as equivalent to the entire world. Aristotle appears to opt for heaven as the totality of the world and then argues that "there is no place or void or time outside the heaven" (Aristotle 1984, 462). Aristotle's desire to cite the opinions of others is also illustrated in his *Meteorology*, where he declares: "Let us now explain the origin, cause, and nature of the milky way. And here too let us begin by discussing the statements of others on the subject" (Aristotle 1984, 564). Other examples are readily available, but one cannot doubt that Aristotle regarded the citation of other opinions as an essential aspect of doing natural philosophy.

Logic was an exception to Aristotle's usual procedure. He did not begin those treatises, especially the *Prior Analytics*, with a summary of previous opinions because there were no previous opinions, and he owed no debt. As Aristotle explains in his *Sophistical Refutations*: "it was not the case that part of the work had been thoroughly done before, while part had not. Nothing existed at all" (Aristotle 1984, 314).

It was also characteristic of Aristotle's style in natural philosophy to inform his readers or listeners about the procedure he intended to follow. Thus, at the outset of *On Interpretation* (*De interpretatione*), Aristotle explains, "First we must settle what a name is and what a verb is, and then what a negation, an affirmation, a statement and a sentence are" (Aristotle 1984, 25) and then proceeds to consider each of these entities. He begins the *Prior Analytics* by announcing the topics he will consider: "First," he says, "we must state the subject of the enquiry and what it is about: the subject is demonstration, and it is about demonstrative understanding. Next we must determine what a proposition is, what a term is, and what a deduction is (and what sort of deduction is perfect and what imperfect); and after that, what it is for one thing to be or not be in another as a whole, and what we mean by being predicated of every or of no" (Aristotle 1984, 39).

From my earlier mention of Aristotle's contributions to biology, it is apparent that he engaged in activities that involved careful observation of the behavior of numerous animals. In some instances, this observational knowledge derived from dissections he did, but often it came from his own desire to observe and report. In his *History of Animals*, Aristotle describes the embryological development of the chicks of the common hen. Here we see him at his best. As will be evident from the passage quoted below, Aristotle obviously broke open eggs that had been laid at the same time and observed the status of the chick at different stages of its development. He tells us that

With the common hen after three days and three nights there is the first indication of the embryo; with larger birds the interval being longer, with smaller birds shorter. Meanwhile the yolk comes into being, rising towards the sharp end, where the primal element of the egg is situated and where the egg gets hatched; and the heart appears, like a speck of blood, in the white of the egg. This point beats and moves as though endowed with life, and from it, as it grows, two vein ducts with blood in them trend in a convoluted course towards each of the two circumjacent integuments; and a membrane carrying bloody fibres now envelops the white, leading off from the vein-ducts. A little afterwards the body is differentiated, at first very small and white. The head is clearly distinguished, and in it the eyes, swollen out to a great extent. This condition lasts on for a good while, as it is only by degrees that they diminish in size and contract. At the outset the under portion of the body appears insignificant in comparison with the upper portion. Of the two ducts that lead from the heart, the one proceeds towards the circumjacent integument, and the other, like a navel-string, towards the yolk. The origin of the chick is in the white of the egg, and the nutriment comes through the navel-string out of the yolk.

When the egg is now ten days old the chick and all its parts are distinctly visible. . . . [After providing much more of a detailed description of the tenth day, Aristotle declares:]

About the twentieth day, if you open the egg and touch the chick, it moves inside and chirps; and it is already coming to be covered with down, when, after the twentieth day is past, the chick begins to break the shell. The head is situated over the right leg close to the flank, and the wing is placed over the head; and about this time is plain to be seen the membrane resembling an after-birth that comes next after the outermost membrane of the shell, into which membrane the one of the navel-strings was described as leading (and the chick in its entirety is now within it), and so also is the other membrane resembling an after-birth, namely that surrounding the yolk, into which the second navel-string was described as leading; and both of them were described as being connected with the heart and the big vein. At this time the navel-string that leads to the outer after-birth collapses and becomes detached from the chick, and the membrane that leads into the yolk is fastened on to the thin gut of the creature, and by this time a considerable amount of the yolk is inside the chick and a yellow sediment is inside its stomach. About this time it discharges residuum in the direction of the outer after-birth, and has residuum inside its stomach; and the outer residuum is white and there comes a white substance inside. By and by the yolk, diminishing gradually in size, at length becomes entirely used up and comprehended within the chick (so that, ten days after hatching, if you cut open the chick, a small remnant of the yolk is still left in connexion with the gut), but it is detached from the navel, and there is nothing in the interval between, but it has been used up entirely. During the period above referred to the chick sleeps, but if it is moved it wakes, looks up and chirps; and the heart and the navel together palpitate as though the creature were respiring. So much as to generation from the egg in the case of birds. (Aristotle 1984, 883–884)

In this remarkable description of the embryonic development of a chick, Aristotle shows his masterful powers of observation and his ability to record what he saw in a scientific manner. There is an air of detachment and objectivity worthy of a great scientist and natural philosopher. Aristotle's description is far superior to an earlier one that appeared in the Hippocratic treatise *On the Nature of the Child* (Cohen and Drabkin 1948, 424–425). Virtually all of the experiences and observations Aristotle made exhibit these same qualities. Whether observing and recording the behavior of animals based on direct observation, reporting observations made by others, or writing about the nature and operation of the terrestrial and celestial regions of the physical world, based upon gross observation and many theoretical constructions about its essential features, Aristotle retained the same calm and impersonal mode of presentation. And yet, underlying this impersonal, detached style was a deep love of nature in all its manifestations, as we see in this famous introductory passage to his biological treatise *On the Parts of Animals*:

Having already treated of the celestial world as far as our conjectures could reach, we proceed to treat of animals, without omitting, to the best of our ability any member of the kingdom, however ignoble. For if some have no graces to charm the sense, yet nature, which fashioned them, gives amazing pleasure in their study to all who can trace links of causation, and are inclined to philosophy. Indeed, it would be strange if mimic representations of them were attractive, because they disclose the mimetic skill of the painter or sculptor, and the original realities themselves were not more interesting, to all at any rate who have eyes to discern the causes. We therefore must not recoil with childish aversion from the examination of the humbler animals. Every realm of nature is marvellous: and as Heraclitus, when the strangers who came to visit him found him warming himself at the furnace in the kitchen and hesitated to go in, is reported to have bidden them not to be afraid to enter, as even in that kitchen divinities were present, so we should venture on the study of every kind of animal without distaste; for each and all will reveal to us something natural and something beautiful. Absence of haphazard and conduciveness of everything to an end are to be found in nature's works in the highest degree, and the end for which those works are put together and produced is a form of the beautiful. (Aristotle 1984, 1003–1004)

Aristotle's positive attitude toward nature and his own desire to remain a careful and objective observer are exemplified in all his works on natural philosophy. But his methodological approach to nature did not include the use of experiments. It has been suggested that Aristotle would have had no interest in experiments on natural phenomena, because experiments require one to alter the behavior of nature artificially and arbitrarily. By altering the natural environment of the thing that is being investigated, we do not observe its natural behavior, because its natural behavior only occurs under natural conditions. "It is therefore senseless to place a substance under artificial conditions for better observation. . . . Experiment, in short, opens up no new access to the facts, and may succeed only in suppressing them" (Waterlow 1982, 34).

In the chick embryo experiment cited earlier, however, Aristotle did intervene in nature, because he realized that only by breaking eggs on different days could he observe what would otherwise be unobservable, namely the embryonic development of chicks. In at least this one instance, Aristotle showed that he would interfere with nature. Perhaps, we should assume that Aristotle would have intervened in nature whenever he could see direct benefit from the intervention. That he hardly ever did so, however, tells us that either he rarely ever saw the direct benefit, or that if he did, he was not ingenious enough to conjure up appropriate experiments that might shed light on the natural phenomena in which he was interested. It is not at all clear that Aristotle was reluctant to intervene in nature. A more likely conjecture might be that he rarely ever thought he had to, because he was convinced that he could derive solutions to most problems by using a priori and deductive means to contemplate the way things had to be.

With hindsight, we can see that Aristotle was in error in much of what he had to say about the physical world. That is hardly surprising for someone who wrote more than 2,300 years ago. But we cannot judge Aristotle's significance and impact on that basis, for we know all too well that much scientific knowledge that appeared in the nineteenth century, and even in the twentieth century, has been shown to be erroneous or misleading, or will be shown to be such. We must rather judge Aristotle on the way he approached nature; on the way he organized his research; and on the style and manner in which he presented scientific knowledge. From that standpoint, as we saw, he earns and deserves high praise, which he customarily received, and still receives, from all who have had occasion to judge him in the ancient, medieval, and modern worlds. He taught those who read and studied his works what nature is, and how they ought to appreciate and study it. Aristotle did this through the medium of his many treatises. For it is the phenomenon of Aristotelianism that clustered around Aristotle's works and thoughts that made his name the dominant force in natural philosophy from late antiquity to the seventeenth century. We must now describe this momentous and extraordinary story.

Chapter 3

Science and Natural Philosophy in
the Roman Empire

THE PRE-SOCRATIC NATURAL PHILOSOPHERS

The science of the Roman Empire was but a continuation of the Greek science that began with a group known as the pre-Socratic natural philosophers in the sixth century B.C. The individuals who comprise this group were active in the period from around 600 B.C. to the time just before Socrates, sometime around 400 B.C. There can be little doubt that the emergence of the pre-Socratic natural philosophers in the Greek city-states along the coast of Asia Minor in the sixth century B.C. marks the beginning of the critical and analytic spirit that would become characteristic of rationalistic Greek science and philosophy. Unfortunately, their works have not survived. Only fragmentary parts were preserved by later authors, many of whom were writing centuries later.

The fragments, however, reveal substantive aspects of the thought of these early Greek natural philosophers. From these fragments, we learn that the pre-Socratics abandoned the supernatural and magical explanations that had been routinely employed in early Greek society and in ancient Egypt and Mesopotamia. Commencing in the Ionian city of Miletus, the first group of pre-Socratic thinkers was known as the Milesians, among whom the most famous were Thales (c. 625–c. 547 B.C.), Anaximander (c. 610–c. 545 B.C.), and Anaximenes (fl. c. 545 B.C.). They, and others who followed, were known as monists, because they believed that the world was composed of a single substance. The varied objects and things we see in the world are but different manifestations of the same substance, which for Thales was water; for Anaximander, an indeterminate entity, or *apeiron*, from which all things emerged and to which they returned; and for Anaximenes, air was the underlying substance of all things. Pythagoras of Samos (c. 560–c. 480 B.C.) was also a monist, but he, and his followers, made number the basic stuff of the universe. They viewed the world as made up of numbers, in some sense, and therefore essentially mathematical in its structure.

The Milesians believed that change did occur and that we can observe those changes. Parmenides of Elea (c. 515–c. 450 B.C.) and his famous disciple, Zeno of Elea (c. 490–c. 425 B.C.), challenged this seemingly self-evident view. Parmenides proposed a radical interpretation. He regarded the human senses as essentially unreliable, and, therefore, did not trust them. He placed his complete trust in reason, demonstrating this by uncompromisingly following the logic of an argument. In a philosophical poem that Parmenides wrote and of which a fair portion has survived, we see that he divided his poem into two parts: The Way of Truth and the Way of Seeming. In the Way of Truth, Parmenides offers a logical argument that being, which is ungenerated, homogeneous, unchanging, motionless, and indestructible, is all that exists. The changes we think we observe are mere illusions.

Zeno of Elea sought to buttress his master's position by formulating a series of paradoxes to demonstrate the impossibility and absurdity of motion. His most famous argument involves Achilles and the tortoise. Zeno argued that if the tortoise is given an initial lead over Achilles, Achilles, the fastest being in the world, cannot overtake the tortoise, the slowest creature in the world. This follows because every time Achilles reaches a point where the tortoise has been, the tortoise has ambled on a bit. Although the distance between them will continually diminish, Achilles will never catch the tortoise. Another version of this argument is one in which Achilles does not even begin to pursue the tortoise. This follows from the assumption that before Achilles can begin to pursue the tortoise, he must first traverse half the distance that separates him from the tortoise; and before he can traverse half the distance, he must traverse one-quarter of the distance; and before he can go one-quarter of the distance, he must go one-eighth; and so on ad infinitum. Thus, Achilles will never begin his pursuit of the tortoise. Zeno also presented a series of paradoxes to show that a plurality of things is impossible, thus defending Parmenides' position that only one thing exists in the world and nothing else can come into being. The arguments of Parmenides and Zeno were deductive in form and involved logical reasoning. They laid the basis for the subsequent Greek emphasis on logic, which culminated with Aristotle's invention of syllogistic, formal logic.

Those pre-Socratics who followed Parmenides had to take account of his denial of the possibility of change. They were not, however, prepared to deny the evidence of their senses: they were convinced that changes of all kinds occurred, but sought to explain observed changes by combinations of unchanging substances. Thus, Empedocles of Acragas (c. 492–c. 432 B.C.) assumed the existence of four elements, or roots, as he called them, namely earth, water, air, and fire, which, in Aristotle's natural philosophy, became the four elements that formed the basis of physics for almost 2,000 years. Empedocles believed the four elements were eternal—that they always existed and would continue to exist into an infinite future. Change occurred either by the coming together of two or more of the four elements, or by their separation. To form the myriad of things in existence, the elements combined in different proportions. As the mechanism of change, Empedocles assumed the existence of two opposing forces, which he called Love and Strife. Love brought elements together and Strife caused their separation. Empedocles believed he was faithful to Parmenides' dictum that no new substances could come into existence from previous material substances. For Empedocles, the unchanging, eternal four elements produce the different things we observe by simple combinations and separations in which the elements remain unchanged.

Anaxagoras of Clazomenae (c. 500–c. 428 B.C.) adopted a very different strategy even as he, like Empedocles, sought to explain change without the emergence of new substances. To avoid the possibility of a new substance coming into being, Anaxagoras assumed that every material thing has a portion of everything else in it. For example, an apple contains hair, wheat, iron, flesh, blood, and a bit of every other existent substance. One substance can change into something else because that into which it changes is already present. Thus, nothing new comes into being, because it already exists within the body in which it appears.

The best-known system of the world presented by pre-Socratic natural philosophers is the atomic theory, first proposed by Leucippus of Miletus (fl. fifth century B.C.) and developed further by Democritus of

Abdera (fl. late fifth century B.C.). To account for the material world and the changes that occur in it, Leucippus and Democritus assumed the real existence of only two things: atoms and the void. The atoms were assumed to exist in all sizes and shapes and to be infinite in number, spread out in an infinite void space and in perpetual motion. Atoms were indivisible, but each atom could combine with other atoms to form the existent bodies of our world. Bodies of all kinds were formed when atoms came together and were juxtaposed to each other; those same bodies would pass away when the atoms dissociated themselves and passed into the void to eventually combine with other atoms to form another body. The qualities we observe and perceive—for example, colors, tastes, touch—are not real but are secondary effects derived from the sizes, shapes, and configurations of the atoms. Because the void is infinite and the number of atoms is infinite, Democritus assumed the simultaneous existence of innumerable worlds—some in process of dissolution while others were always coming into existence. In the history of cosmological theory, Leucippus and Democritus were probably the first to assume the existence of innumerable worlds, perhaps even an infinity of worlds.

THE EMERGENCE AND DEVELOPMENT OF THE SCIENCES IN THE GREEK WORLD

By their naturalistic and largely rationalistic approach to the natural world, the pre-Socratic natural philosophers established a foundation for the great advances in Greek science that followed. Indeed, the first of these sciences—medicine—developed contemporaneously with pre-Socratic thought.

THE LIFE SCIENCES

Medicine

In its earliest stages, Greek medicine was practiced in the temples of Asclepius, where the priests of Asclepius sought to effect cures by interpreting the dreams of those who came for treatment. Temple medicine continued for some time. By the fifth century B.C., medical schools were in existence on the island of Cos, in Cnidus, and in southern Italy. The most famous of these schools was in Cos, associated with the name of Hippocrates of Cos. Medical schools in this period had the nature of secret brotherhoods. The famous Hippocratic oath suggests this, because each member of the school swore to teach the medical art only to his sons, the sons of his masters, and to duly apprenticed pupils.

Hippocrates of Cos was born around 460 B.C. and died around 370 B.C. Mentioned by Plato and Aristotle, he acquired enormous prestige in the generation after his death. Approximately seventy medical treatises are attributed to him, although members of his school probably wrote many of them. As a result, the works are rather divergent and their viewpoints are sometimes in conflict. The Hippocratic treatises vary in their emphases: some stress theory, while others are empirical, staying close to facts and customs; some of the works balance experience and theory. Hippocratic medicine was probably no more effective than temple medicine. But the school of Hippocrates laid the basis for a rational approach to medicine. We can see this in one of the Hippocratic treatises, *On the Sacred Disease*, which referred to epilepsy. The author declares that "this disease [epilepsy] is in my opinion no more divine than any other; it has the same nature as other diseases, and the cause that gives rise to individual diseases. It is also curable, no less than other illnesses, unless by long lapse of time it be so ingrained as to be more powerful than the remedies that are applied" (Cohen and Drabkin 1948, 473–474). Although many in the ancient world regarded epileptic seizures as caused by divine intervention, the Hippocratic author of *On the Sacred Disease* regarded it as a disease like any other, to be treated by natural and rational methods.

Because no licenses were required to practice medicine in ancient Greece, many charlatans pretended to be physicians, traveling from town to town and practicing medicine on unsuspecting townspeople. Medical ethics were virtually unknown. It was with the hope of remedying this dangerous situation that physicians of the Hippocratic School formulated the famous Hippocratic Oath. Because it is relatively brief, I shall cite the whole of it:

I swear by Apollo Physician and Asclepius and Hygieia and Panaceia and all the gods and goddesses, making them my witnesses, that I will fulfill according to my ability and judgment this oath and covenant:

To hold him who has taught me this art as equal to my parents and to live my life in partnership with him, and if he is in need of money to give him a share of mine, and to regard his offspring as equal to my brothers in male lineage and to teach them this art—if they desire to learn it—without fee and covenant; to give a share of precepts and oral instruction and all the other learning to my sons and to the sons of him who has instructed me and to pupils who have signed the covenant and have taken an oath according to the medical law, but to no one else.

I will apply dietetic measures for the benefit of the sick according to my ability and judgment; I will keep them from harm and injustice.

I will neither give a deadly drug to anybody if asked for it, nor will I make a suggestion to this effect. Similarly I will not give to a woman an abortive remedy. In purity and holiness I will guard my life and my art.

I will not use the knife, not even on sufferers from stone, but will withdraw in favor of such men as are engaged in this work.

Whatever houses I may visit, I will come for the benefit of the sick, remaining free of all intentional injustice, of all mischief and in particular of sexual relations with both female and male persons, be they free or slaves.

What I may see or hear in the course of the treatment or even outside of the treatment in regard to the life of men, which on no account one must spread abroad, I will keep to myself holding such things shameful to be spoken about.

If I fulfill this oath and do not violate it, may it be granted to me to enjoy life and art, being honored with fame among all men for all time to come; if I transgress it and swear falsely, may the opposite of all this be my lot. (Edelstein 1943, 3)

This is a laudable document. High standards were expected from the physicians who took this oath. The large number of medical treatises that have been preserved from the school of Hippocrates reveal a generally high level of medical achievement. One of the great tenets of the Hippocratic doctors was the healing power of nature. The key idea was to work with nature, but to let it do the healing, if at all possible. Only when nature is not adequate to the task is the physician to intervene and aid nature to the extent that is necessary. At a time when little was known about the workings of the human body, this was a wise and sensible approach.

The Hippocratic doctors believed that the body was healthy when its basic fluids were in equilibrium and harmony. These four fluids were blood, phlegm, yellow bile, and black bile, as we learn from the Hippocratic treatise *On the Nature of Man.* "The body of man," we are told,

has in itself blood, phlegm, yellow bile, and black bile; these make up the nature of his body, and through these he feels pain or enjoys health. Now he enjoys the most perfect health when these elements are duly proportioned to one another in respect of compounding, power and bulk, and when they are perfectly mingled. Pain is felt when one of these elements is in defect or excess, or is isolated in the body without being compounded with all the others. (Cohen and Drabkin 1948, 488)

The four humors became one of the foundations of Greek medicine and served all the physicians who followed the Greek medical tradition.

Among the greatest glories of Hippocratic medicine were the clinical observations of the course of various diseases that were given in the treatises called *Epidemics*. These are usually day-by-day accounts of the progress of a disease or ailment. They are succinct and vivid, with most ending in death. Another important feature of the Hippocratic treatises is the debate between those physicians who believed that philosophy was important in medicine and those who did not. From the history of Greek medicine we learn that those who regarded philosophy as an important tool in the study of medicine helped shape that discipline. Many aspects of Greek medicine reveal the impact of philosophy. It was the philosophical spirit that prompted Greek physicians to determine the course of a disease, to distinguish different types of diseases, to investigate the causes of various diseases, to produce theories of the prognosis of diseases, and to formulate methodological procedures for the treatment of ailments.

Following the conquests of Alexander the Great and the spread of Greek culture and science into Egypt, much of the Near East, and as far east as India, Greek science flourished in the Hellenistic period, from 300 B.C. to the beginning of the Roman Empire and the birth of Christ. It was in the Hellenistic period, in the city of Alexandria, Egypt, that the Greeks developed a number of medical schools, each of which had its own distinctive approach and methodology. The Empiricist school emphasized experience and rejected theoretical medicine based on reasoned argument. The Dogmatist school emphasized reason but also regarded direct observation of the internal organs as vital to a proper understanding of the ailments that afflict the human body. To enable physicians to become familiar with those organs, the Dogmatists became the first physicians in history to dissect and even vivisect the human body.

The beginnings of the process by which human dissection became a part of Greek medicine is traceable to Plato's description of the death of Socrates in the dialogue Plato called *Phaedo*. Up to Plato's time, the Greeks viewed the dead human body as sacred, and they firmly believed that it had to be properly buried, fearing that if this duty was neglected, the corpse would take vengeance on those relatives who had abandoned their solemn responsibilities. Hence, dissection would have been regarded as the mutilation of a dead body and was, therefore, considered unthinkable. The Greeks also placed great emphasis on the "body beautiful" and always thought in terms of preserving the body rather than mutilating it.

Plato's attitude was radically different. He regarded the human soul as far more important than the body: the soul was immortal, whereas the body was perishable. Plato urged his readers to look after the well-being of their souls and to forget about their bodies, as he reveals in an account of Socrates' death in the *Phaedo*. In 399 B.C., Socrates was tried for impiety in Athens and found guilty. His punishment was to drink poison in the form of a cup of hemlock. The *Phaedo* reports the alleged discussions that Socrates had with his friends on the day he drank the poison at sunset and ended his life. When the time arrives for Socrates to drink the hemlock, one of his friends, Crito, asks, "And how are we to bury you?" Socrates replies: " 'Anyhow you like, if you can catch me, and I don't elude you' " (Plato 1955, 139). Crito thinks in terms of a traditional burial, and Socrates complains that he has failed to convince Crito "that when I drink the poison I shall no longer remain with you, but take my leave of you and go off to some 'joys of the blessed' " (Plato 1955, 139–140). Socrates informs Crito that he can bury him any way he likes. After Socrates drinks the poison and lies near death, he utters these final words: "Crito, we owe a cock to Asclepius; please pay it—do not neglect it" (Plato 1955, 143). The custom was to offer a cock to Asclepius, the god of healing, after a cure had been effected. Socrates believed that his death was really a cure from the ills of human existence. The cure lay in his firm belief that with death his soul would be released for an eternal life without the hindrance of a body.

Thus did Plato use a philosophical approach to overcome the traditional fear of the dead body. However, Plato's greatest student, Aristotle, who dissected numerous animals, did not try to break the long-standing Greek taboo against dissecting the human body, as we learn from Aristotle's *History of Animals*, in which he explains that while we know the arrangement of the external parts of the human body, "this is not the case with the inner parts. For the fact is that the inner parts of man are to a very great extent unknown, and the consequence is that we must have recourse to an examination of the inner parts of other animals, whose nature in any way resembles that of man" (Aristotle 1984, 788). Plato's attitude toward the dead human body proved influential outside of Greece when physicians of the Dogmatist school in Alexandria not only overcame fear of the dead body but, in the course of the third century B.C., also took the dramatic step of dissecting it and perhaps even vivisecting it.

Writing in Rome in the first century A.D., Aulus Cornelius Celsus (fl. c. A.D. 25) composed an important treatise titled *On Medicine* (*De medicina*). In the introduction, Celsus writes about the Dogmatist physicians and declares:

Moreover, as pains, and also various kinds of diseases, arise in the more internal parts, they hold that no one can apply remedies for these who is ignorant about the parts themselves; hence it becomes necessary to lay open the bodies of the dead and to scrutinize their viscera and intestines. They hold that Herophilus and Erastistratus did this in the best way by far, when they laid open men whilst alive—criminals received out of prison from the kings— and whilst these were still breathing, observed parts which beforehand nature had concealed, their position, colour, shape, size, arrangement, hardness, softness, smoothness, relation, processes and depressions of each, and whether any part is inserted into or is received into another. (Cohen and Drabkin 1948, 471–472)

Celsus' charge that the two most famous Dogmatist physicians in Alexandria, Herophilus of Chalcedon (b. c. 330 B.C.) and Erasistratus of Iulis (b. c. 304 B.C.), practiced vivisection has not been conclusively demonstrated, although it is likely they did so. Unfortunately, neither of these Dogmatist physicians left any extant treatises in which they discuss their work. Our knowledge of their contributions is derived from later writers, especially Galen, the great physician of the second century A.D. Nevertheless, G.E.R. Lloyd has plausibly conjectured: "When we reflect that the ancients regularly tortured slaves in public in the law courts in order to extract evidence from them, and that Galen, for example, records cases where new poisons were tried out on convicts to test their effects, it is not too difficult to believe that the Ptolemies permitted vivisection to be practised on condemned criminals" (Lloyd 1973, 77). Celsus, and later Christian authors such as Tertullian and Saint Augustine, reproached the Dogmatists for cruelty. Despite his denunciation of vivisection, Celsus approved of dissection, regarding it as essential for students of medicine.

By their human dissections, Herophilus and Erasistratus laid the basis for the sciences of anatomy and physiology. They usually dissected the bodies of executed criminals whose bodies were handed over to them on the authority of the Ptolemaic kings of Egypt. The knowledge they derived from their dissections is impressive. Herophilus showed that it is the brain that is the center of the nervous system and the seat of intelligence, and not the heart as Aristotle had falsely believed. He identified and distinguished the cerebrum and cerebellum in the brain, distinguished between arteries and veins, described the optic nerves and retina in the eye, and made numerous other contributions. He also named a number of organs, including the duodenum. Erasistratus added to our knowledge of anatomy, too, describing the function of the epiglottis and distinguishing between motor and sensory nerves.

Although the ancient Egyptians had also cut into the human body in order to embalm dead pharaohs, as well as members of the higher nobility, these cuttings were for religious purposes, to preserve dead pharaohs in an eternal afterlife. The objectives of the Greek physicians differed radically from their Egyptian predecessors. They dissected human cadavers for the sole purpose of learning about the internal organs of the human body. The Greeks used dissection, and perhaps vivisection, for scientific purposes, whereas the ancient Egyptians used it for religious purposes. Unfortunately, the vivisections of prisoners, who were regarded as worthless and expendable, are deplorable even if they provided useful medical information. It reminds us too vividly of medical experiments performed by Nazi physicians on concentration camp victims and prisoners of war in World War II during the last century.

Numerous other Greek physicians contributed to medical history in the Hellenistic period, encompassing the last three centuries B.C. Although Rome was the dominant force in Hellenistic times, medicine, like all other sciences, was largely a Greek enterprise.

Biology

Some, if not much, that Herophilus and Erasistratus discovered was also relevant to biology. But they focused on the human body and the functioning of its parts. The larger issues in biology, such as a concern for the classification of animals into species and genera and the description of a great variety of animals, were almost exclusively the work of Aristotle, who single-handedly established the science of biology. Aristotle made the most significant contributions to biology in the ancient world (in chapter 2, we saw that he dissected numerous animals and was a keen observer of animal behavior).

Aristotle began the long historical process of the classification of animals. He regarded species as permanent and unchangeable and was therefore not an evolutionist. In his view, one species could not evolve over time to become another species; rather, living things form a ladder of nature. They proceed little by little from inanimate matter to plants, which have a vegetable soul that enables them to obtain nutriments and to reproduce. Plants, in turn, ascend by small degrees toward the animal world. Animals have a level of vitality that enables them to obtain food, to have sensations, and to move about (locomotion). So subtle is the ascent of life forms that Aristotle was sometimes unable to determine whether some living things in the ocean were plants or animals. The final living form in the ladder of nature is man, who, according to Aristotle, has a rational soul that enables him to engage in rational discourse, an activity in which plants and animals cannot participate.

Aristotle considered many of the fundamental problems of biology. Besides classification, he studied reproduction and the processes of nutrition. Indeed, his classifications were largely based on different modes of reproduction. His extraordinary powers of observation have evoked the admiration of modern biologists. He knew that cetaceans (whales, porpoises, dolphins) were mammals and realized that they therefore had lungs, breathed air, were viviparous (i.e., their offspring were born alive), and suckled their young. Despite the fact that cetaceans lived in water, Aristotle correctly classified them as mammals rather than fish. Until the eighteenth and nineteenth centuries, biology was overwhelmingly the biology of Aristotle. There were no rivals.

THE EXACT SCIENCES

Mathematics

Both the ancient Babylonians and the Egyptians contributed to the beginnings of geometry, with the Babylonians reaching a high degree of proficiency in algebra. The Greeks were indebted to these two civilizations for some of their mathematical knowledge and understanding. Neither the Egyptians nor the Babylonians, however, formulated a concept of mathematical proof, a crucial contribution that was first made by the Greeks, perhaps by Thales in the first half of the sixth century B.C. Although the attribution to Thales has been contested, there is, nonetheless, ample evidence that the Greeks were keenly interested in mathematics since the sixth century B.C. and that, in the course of the fifth century B.C., they introduced the concept of rigorous mathematical proof. Pythagoras of Samos and his followers, the Pythagoreans, were famous for their mystical interest in numbers, number theory, and arithmetic. They were also interested in geometry, and the Pythagorean theorem is one of the legacies attributed to Pythagoras, although it has been shown that the Babylonians at the time of King Hammurabi also knew the Pythagorean theorem, and it is possible that Pythagoras learned it from them. Pythagoras, or his followers, however, proved the theorem. Another of their noteworthy contributions was the discovery of incommensurability: they realized that the diagonal of a square is incommensurable to its side. The Pythagoreans regarded this discovery as a blow to their theory of proportion, which could only be applied to commensurable magnitudes that had a common measure, but not to incommensurable magnitudes.

The best-known and most explicit example of mathematical rigor and proof in Greek geometry is Euclid's *Elements* (in thirteen books), which is also the first complete Greek mathematical treatise to have survived intact. Its enormous utility and ubiquitous use guaranteed its preservation to the present day. The Greeks had already written books of elements long before Euclid composed his treatise. Great strides in geometry had been made at Plato's Academy, where a number of great Greek mathematicians had come to study with Plato, who was keenly interested in mathematics. The most notable were Eudoxus of Cnidus (c. 400–c. 347 B.C.) and Theaetetus (c. 417–369 B.C.). Our knowledge of the history of Greek mathematics is largely derived from later accounts, the most important of which is by Proclus (A.D. 410–485), a neo-Platonic philosopher, who incorporated much historical information into his *Commentary on the First Book of Euclid's "Elements"*. The value of Proclus' account lies in the often unique information he furnished about Euclid's predecessors and successors.

Euclid (fl. c. 295 B.C.) lived and worked in Alexandria. Virtually nothing more is known about his life. Proclus informs us that Euclid "put together the elements, arranging in order many of Eudoxus' theorems, perfecting many of Theaetetus', and also bringing to irrefutable demonstration the things which had been only loosely proved by his predecessors" (Bulmer-Thomas 1971, 414). Proclus also reports a story that has become famous, although its truth cannot be verified. King Ptolemy of Egypt "once asked him if there were a shorter way to the study of geometry than the *Elements*, to which he replied that there was no royal road to geometry" (Bulmer-Thomas 1971, 414).

The *Elements* is a synthetic mathematical treatise in which Euclid always proceeds from what is known to something unknown. Occasionally, he resorts to *reductio ad absurdum* proofs—that is, proofs that involve a reduction to absurdity—in which Euclid shows that if you do not accept a certain conclusion, impossible consequences follow. In addition to propositions, Euclid's *Elements* include definitions, postulates, and axioms. The postulates and axioms are placed just before the beginning of Book I and form the basis of Euclid's geometry. The postulates are relevant only to geometry, whereas the axioms, or "common notions," as Euclid called them, are relevant to all demonstrative sciences. Postulates

are unproved but accepted premises. The five postulates of the first book are the most noteworthy, since they really define the whole character of the space and geometry under consideration. The first permits us to draw a straight line from any one point to any other point; the second allows us to extend that line indefinitely from either extremity; the third allows us to draw a circle from any center at any distance from the center; the fourth asserts that all right angles are equal; and the fifth, which is the famous parallel postulate, tells us under what conditions straight lines will intersect. The genius of Euclid in recognizing this fifth postulate as a postulate and not attempting formal proof has often been remarked. (Clagett 1957, 59)

The axioms, or common notions, are self-evident and need no explanation. Because they are few and brief, and because most students will probably recognize them, it is simpler to cite the five axioms than to elaborate further:

1. Things which are equal to the same thing are also equal to one another.
2. If equals be added to equals, the wholes are equal.
3. If equals be subtracted from equals, the remainders are equal.
4. Things which coincide with one another are equal to one another.
5. The whole is greater than the part. (Euclid 1956, 1:155)

With the exception of a few primary geometrical entities, such as points and lines, the existence of which cannot be proved but must be assumed, definitions in geometry define what things are but do not claim that they exist. Existence must be proved. To illustrate Euclidean definitions, the following are drawn from the twenty-three definitions of the first book:

1. A point is that which has no part.
2. A line is breadthless length.
3. The extremities of a line are points.
11. An obtuse angle is an angle greater than a right angle.
12. An acute angle is an angle less than a right angle.
13. A boundary is that which is an extremity of anything.
23. Parallel straight lines are straight lines which, being in the same plane and being produced indefinitely in both directions, do not meet one another in either direction. (Euclid 1956, 1:153, 154)

Euclid includes two kinds of propositions: constructional propositions and straight theorems. As the name implies, a constructional proposition involves the construction of a figure. Indeed, Euclid begins the *Elements* with the construction of an equilateral triangle on a finite straight line. Book I, Proposition 1 reads: "On a given finite straight line to construct an equilateral triangle" (Euclid 1956, 1:241). Most propositions, however, are theorems in which Euclid announces some geometric principle and then proves it, as, for example, in Book I, Proposition 15: "If two straight lines cut one another, they make the vertical angles equal to one another" (Euclid 1956, 1:277).

By the time Euclid composed his *Elements*, Greek geometers had formalized the way geometrical propositions were presented to students and readers. There were six basic parts to a typical theorem:

1. *The enunciation*, which states what is given and what is to be proved. Thus, in Book I, Proposition 27, the enunciation declares: "If a straight line falling on two straight lines make the alternative angles equal to one another, the straight lines will be parallel to one another" (Euclid 1956, 1:307).
2. The next element in a theorem is the *setting out*. These are the opening words of Proposition 27, which assert: "For let the straight line *EF* falling on the two straight lines *AB, CD* make the alternate angles *AEF, EFD* equal to one another" (see Figure 3.1) (Euclid 1956, 1:307).

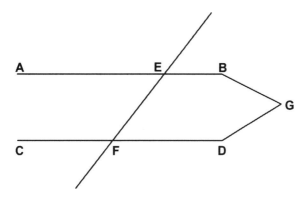

Figure 3.1. Book 1, Proposition 27 of Euclid's *Elements*, which proves that if angle AEF is equal to angle EFD, then AB is parallel to CD. From Euclid 1956, 1: 307.

3. The third element is the *definition* or *specification*, which clearly and distinctly asserts what is sought. Thus in Proposition 27, the *definition* declares: "I say that *AB* is parallel to *CD*" (Euclid 1956, 1:307).

4. The fourth step was called the *construction* or *machinery*, which adds data that are needed to find what is sought, but which is lacking. The very next lines of the proof constitute the *construction*: "For, if not, *AB, CD* when produced will meet either in the direction of *B, D* or towards *A, C*" (Euclid 1956, 1:307).

5. The fifth step in a proposition is the *proof* itself, which draws the necessary inferences by reasoning deductively from the acknowledged facts.

6. The final stage of a typical demonstration is the *conclusion*, which reiterates what has been demonstrated. Here Euclid simply repeats the enunciation, or the first step (above). This is done by the words "Therefore, etc.," where the one simply substitutes the enunciation for "etc." This is usually followed on the next line by the letters "Q.E.D.," which stand for the Latin words "quod erat demonstrandum," or, "that which was to be demonstrated" (Euclid 1956, 1:308).

Of these six basic elements of a Euclidean theorem, three—enunciation, proof, and conclusion—appear in all theorems. The others are introduced only as needed by the requirements of the demonstration.

Euclid's *Elements* formed the basis of Greek geometry. It was a model of deductive logic and mathematical synthesis. Utilizing the solid base of Euclidean geometry, Greek mathematicians reached a very high level of achievement. The two most important mathematicians who contributed to the advancement of Greek mathematics in the Hellenistic period were Archimedes (c. 287–212 B.C.) and Apollonius of Perga (fl. c. 200 B.C.).

Archimedes was unquestionably the greatest mathematician of the ancient world. He was not only a mathematical genius, but was also renowned as a great engineer who devised various pulley-like devices by means of which a small force could move a very large weight. As the ultimate example of a small force moving a large weight, Archimedes is reported to have exclaimed: "Give me a place to stand on, and I will move the earth" (Clagett 1970, 1:213). Numerous mathematical works are attributed to him, some extant and some no longer existent. Archimedes used geometrical analysis to solve problems in statics and hydrostatics, as in his famous treatises *On the Equilibrium of Planes* and *On Floating Bodies*. Archimedes tackled some monumental problems. For example, in *The Sand-Reckoner*, he attempted to measure the number of grains of sand in the universe after making certain assumptions about the size of the world. In modern notation, Archimedes concludes that the number of grains of sand in the universe is not larger than 10^{63}. To represent such a number, Archimedes had to invent a special system to express large numbers.

In the third proposition of *On the Measurement of the Circle*, Archimedes calculates the ratio of the circumference of a circle to its diameter, a value that we call pi π, a term that is of modern vintage and was not used by the Greeks. As his approximation for the value of π, Archimedes gives $3 \, 1/7 > \pi > 3 \, 10/71$. In *On the Equilibrium of Planes*, Archimedes proves the law of the lever by means of geometry. In that same treatise, Archimedes was the first to bring together the elementary theorems of statics. From Euclid, Archimedes further developed the method of exhaustion to measure the area of a circle. In the application of the method of exhaustion, the area of a circle is calculated by inscribing or circumscribing polygons inside or outside of the circle. As the sides of the polygon are doubled, the area of the polygon, which is always known, becomes larger and larger if it is inscribed, or smaller and smaller if it is circumscribed. Thus, whether it is inscribed or circumscribed, the polygon approaches the area of the circle. The difference between the circle and the polygon can be made as small as one desires, and thus a good approximation of the area of a circle can be determined. Archimedes was undoubtedly a genius in mathematics, and his influence on the subsequent history of mathematics was deep and lasting in both the Middle Ages and Renaissance.

Although numerous mathematical treatises have been attributed to Apollonius of Perga (fl. c. 200 B.C.), only two have survived, one of which is of momentous importance, the *Treatise on Conic Sections*. In this work, Apollonius gives a thorough treatment of conic sections, to which he gave the names of ellipse, parabola, and hyperbola. His treatise, along with the works of Archimedes, would prove crucial in the pure mathematics of the Renaissance, as well as in the application of mathematics to physics, astronomy, and mechanics.

Astronomy

In its early pre-Socratic phase, Greek astronomy was based on broad cosmological descriptions of the celestial region or on the nature of our earth: what is its shape, position, source of support, and relation to other celestial bodies. Thales, for example believed that the earth remains in one position because it is supported by water, whereas Anaximander argued that the earth remains in the center of everything because it is equidistant from everything and has no inclination to move in one direction than in another; it therefore remains motionless in the center of the world. Anaximenes believed that the earth rests on air, which he regarded as the basic substance out of which all things were made. The Pythagoreans regarded the earth as a sphere but, in a radical departure from their predecessors and contemporaries, removed the earth from its central position and located it among the celestial bodies. Like all celestial bodies, the earth was said to move in a circle around a central fire, which was regarded as the hearth of the universe. The central fire is not observable from our earth, because a counter-earth, which also moves around the central fire, lies between the two. We cannot see the counter-earth because we live on the hemisphere that faces away from it. The Pythagoreans also insisted on the divinity of the celestial region and of all its planets and stars. With the possible exception of the Greek atomists, the divinity of the celestial bodies became a standard feature of ancient Greek astronomical thought.

In this early period of the fifth century B.C., astronomical observations were made and data compiled that were used to construct a calendar. In the latter part of the fifth century B.C., a body of astronomical observations and ideas were developed on which subsequent astronomers could draw. It was used to help resolve calendrical problems. The Greeks had a lunar calendar but sought to keep their lunar calendar in step with the solar year. Since the solar year is longer than the lunar year, it was necessary to intercalate months to bring the two into close agreement. Meton and Euctemon were two Athenian astronomers who made their major contributions around 430 B.C., the former suggesting a nineteen-year solar cycle containing 235 lunar months, while the latter recognized the inequality of the four seasons and offered estimates for the length of each (Lloyd 1970, 82).

To convey an idea of how daring and imaginative Greek cosmological thought was in this period, I need only reiterate what I said earlier in this chapter about Leucippus (fl. fifth century B.C.) and Democritus of Abdera (fl. late fifth century B.C.), when I had occasion to mention these two proponents of Greek atomism and an infinite universe and to explain that they believed in the simultaneous existence of many worlds and perhaps even an infinity of worlds. According to their interpretation, we live in a universe in which the atoms of some worlds are in process of dissociating and bringing those worlds to an end, while other atoms are in process of coming together and forming new worlds that will in the far distant future also pass away. This is probably the first time in the history of astronomical and cosmological thought that an infinite universe was proposed with innumerable worlds existing simultaneously within it.

In Greek civilization, it was the pre-Socratic natural philosophers who may rightly be said to have begun the serious study of astronomy and cosmology. These disciplines were significantly advanced with the appearance of Plato's Academy and the mathematicians and astronomers who were drawn to it. Plato sought to deemphasize the gathering of observational data in astronomy and to replace that quest with mathematical astronomy. That is, he sought to represent astronomical motions mathematically, believing that was the true path to genuine knowledge about the celestial motions. Later writers report that Plato posed the following question to students of astronomy: "By the assumption of what uniform and orderly motions can the apparent motions of the planets be accounted for?" (Lloyd 1970, 84). This application of mathematics to account for the apparent motions of the planets "by uniform and orderly motions" came to be called "saving the phenomena." These motions were almost always assumed to be circular. As G.E.R. Lloyd has expressed it, "The problem could be rephrased, then, as being how to combine various uniform circular motions in such a way that their resultant corresponds to the observed movements of the planets" (Lloyd 1970, 85). This was the basic approach that technical astronomers followed from the time of Plato to the seventeenth century.

The earliest effort to represent the seemingly irregular motions of the planets by uniform circular motions began at Plato's Academy with the astronomical system proposed by Eudoxus of Cnidus (c. 400–c. 347 B.C.), one of the greatest mathematicians and astronomers in antiquity, who may have been the first to present the substantive content of Books V, VI, and XII of Euclid's *Elements*. His great contribution to astronomy, however, was to represent the planetary motions by a system of homocentric spheres—that is, a system of concentric circles in which the earth is assumed to lie at the center. By Eudoxus' time, the Greeks had identified four major motions with which each planet moved simultaneously. Eudoxus' objective, and that of all astronomers thereafter, was to construct a mathematical system, based on combinations of circular motions, that would represent the four simultaneous motions and thus locate the planets' celestial positions as accurately as possible. The four motions with which each planet moved are as follows:

1. a daily motion from east to west
2. a motion around the zodiac from west to east
3. stations and retrogradations, that is, each planet seems to stop its forward motion and remain stationary for a certain period and then begin a retrograde motion in the opposite direction for a certain time, before again moving in its regular west-to-east direction
4. changes in latitude, that is, changes in a planet's north-south location, where it is sometimes higher in the sky and at other times lower.

During the Hellenistic period and down to the time of Claudius Ptolemy in the second century A.D., Greek astronomers devised different ways of "saving the phenomena"; that is, they devised different combinations of rotating circles to represent the four basic motions and then, from those motions derived the positional relations of each planet with respect to the earth. At least four major configurations are distinguishable. To illustrate their differences, each will be represented by means of a characteristic diagram.

The earliest of these was Eudoxus' system of homocentric spheres in the fourth century B.C., which is represented in Figure 3.2. Marshall Clagett, from whose book this figure is taken, explains the significance of each of the four spheres as follows:

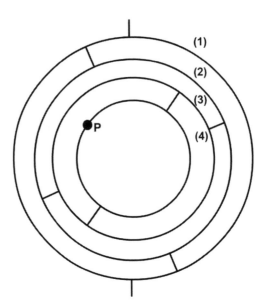

Figure 3.2. Eudoxus' system of homocentric spheres as applied to a single planet. (Clagett 1957, 88.)

Sphere (1), the outermost sphere, rotates from east to west on its axis every twenty-four hours to account for the daily rising and setting of the planet. The poles of sphere (1) lie on a north–south axis. The rotation of sphere (2) accounts for the great encirclement of the planet from west to east through the zodiacal band; its axis is accordingly inclined to that of sphere (1) in about the same angle as the zodiacal band is inclined to the celestial equator (i.e., the equator of the first sphere). The third and fourth spheres rotate in equal times but in opposite directions. Together they account for the looping movement of the planet (i.e., for the stations and retrograde movement . . .) and for some movement in latitude. The poles of the third sphere lie in the zodiacal band (i.e., in the equator of sphere (2)). The axis of the fourth sphere is inclined at an angle to the axis of the third sphere that varies for each planet, just as the speeds of spheres (3) and (4) vary for each planet. The planet (P) is carried on the equator of the fourth sphere. Now the combined movement of the third and fourth spheres will . . . cause the planet to describe about the zodiac the curve called in Greek the hippopede, or "horse-fetter". . . . This curve bears a fair resemblance to the looping motion described by the planets. (Clagett 1957, 88–89)

Eudoxus assumed a total of twenty-seven spheres for all the planets: three for the sun; three for the moon; four for each of the other five planets for a total of twenty; and one sphere for the daily rotation of the heavens. His homocentric system was founded on inadequate observation. Like all other Greek astronomers, Eudoxus did not treat the planetary motions as part of a single celestial system. Rather, he described the motions of each planet as independent of all other planetary spheres. Among some of its major deficiencies, the homocentric system did not allow for variations of planetary distances from the earth, a phenomenon that was discovered later from the observed variability of brightness for each planet, and it failed to explain the inequality of the seasons. Nevertheless, Eudoxus' system remained popular, and a few later astronomers sought to improve it. But Eudoxus and his successors were in agreement that their homocentric spheres were purely mathematical constructions without physical reality. Their efforts were solely intended to account for the positions of the planets, without any claims of actually depicting the physical cosmos.

It was Aristotle who converted Eudoxus' mathematical system into a physical system of the world. Aristotle integrated Eudoxus' separate sets of planetary spheres into a single system of concentric spheres, ranging from the outermost celestial sphere of the fixed stars to the innermost sphere involved in the moon's motion. Although the total number of spheres Aristotle intended to incorporate into his system is in dispute—the number ranges from forty-nine to fifty-five—it is clear that Aristotle assumed that all the spheres were in contact. This was a necessary move, because Aristotle firmly believed that all motions in the world begin with the outermost sphere of the fixed stars and are transmitted successively down to the region below the moon—that is, to the earth where these motions affect all living things.

There were many problems with Aristotle's system of physical concentric spheres, one of the most important being its unsuccessful attempt to explain how the effects of celestial motions could be properly transmitted to the earth by means of the mechanisms he employed. The problem was that if all the spheres were in contact, each sphere would be affected by the motions of all the spheres above it. For example, the motion of the innermost sphere of Saturn's four spheres would impinge on the motion of the outermost sphere of Jupiter's four spheres. To resolve this problem, Aristotle assigned what he called "unrolling" spheres to prevent the motions of a superior planet from affecting the motions of the spheres belonging to the planet immediately below it. Thus, to Saturn's four spheres Aristotle added three unrolling spheres, which moved in the opposite direction of Saturn's second, third, and fourth spheres. By this means, Aristotle sought to cancel the movements of all of Saturn's spheres except the first, or outermost, sphere, which moved with the daily motion. In this way, Jupiter's first, or outermost, sphere would be affected only by the daily motion of Saturn, and thus its daily motion would be intact. The motions of Jupiter's three inner spheres were then cancelled by the action of its three unrolling spheres. The same tactic was applied to all the other planets in descending order.

Aristotle's system was an astronomical failure but a huge physical success. Astronomers soon abandoned Eudoxus' homocentric spheres, but for the next two thousand years, natural philosophers and cosmologists almost unanimously adopted some version of Aristotle's physical system of concentric spheres (see Figure 2.1).

As better observations became available, Greek astronomers moved away from homocentric spheres to other combinations of uniform circular motions. Whatever combinations of circles Greek astronomers used to represent planetary motions, they were agreed that their sole objective was to "save" the astronomical phenomena as accurately as possible, not to discover the real and true configuration of the celestial region. That task fell to the speculations of natural philosophers.

There is no doubt that Greek astronomy reached great heights in the Hellenistic period. Its many achievements, with significant additions, would be incorporated into the climactic work in astronomy by the great Claudius Ptolemy in the second century A.D., a work that was not surpassed until the sixteenth century, when Copernicus published his monumental treatise.

Following the time of Eudoxus, Greek mathematical astronomers proposed at least three major geometrical schemes based on different combinations of uniform, circular motions. All of these are found in Claudius Ptolemy's *Mathematical Syntaxis,* composed in the second century A.D. and unanimously regarded by modern scholars as the greatest surviving astronomical treatise from the ancient world (see Figure 3.3). This great work came to be known as the *Almagest,* which was the title it received when it was translated into Arabic. Two of Ptolemy's three basic models were first put forth in the third century B.C. by Apollonius of Perga, the great mathematician.

The first model was based on an eccentric circle, that is, a circle in which the earth does not lie at the center, so that the planet does not move with uniform motion around the earth, but moves uniformly around a geometric point located at the center. This is illustrated in Figure 3.4, where O is the sun's circular path, S is the sun, E is the

Figure 3.3. Ptolemy observes the stars using a quadrant. From Gregor Reisch's *Margarita philosophica.* (Basel, 1517. Courtesy Lilly Library, Indiana University, Bloomington.)

earth, and *C* is the center of O, the sun's circular path. The advantage of this system was that it could represent the varying speeds of planets; that is, planets are observed to move slower when they are farther from the earth and quicker when closer to the earth.

The second basic astronomical model was the "epicycle on a defer-

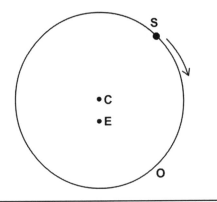

Figure 3.4. Eccentric circle showing the motion of the sun around the earth. In this model, the earth does not lie at the center. (Clagett 1957, 94.)

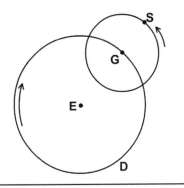

Figure 3.5. Epicycle on a deferent circle showing the motion of the sun around the earth. (Clagett 1957, 94.)

ent." Here the earth was again located in the center, but the planet's varying distances from the earth were now represented by a small circle placed on a large circle around which it rotated. The large circle was called the *deferent*, or the "bearing circle," and the small circle that moved around the circumference of the deferent circle was called an *epicycle*, or a circle on a circle. In Figure 3.5, the earth, E, is at the center of the deferent circle, D; the sun, S, moves around G, the center of the epicycle. The epicycle, in turn, moves around the circumference of D. In this way, the sun's distance from the earth varied just as in the

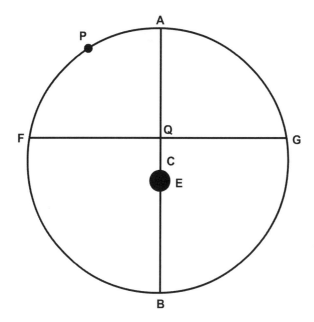

Figure 3.6. The equant model proposed by Ptolemy. (Lindberg 1992, 102.)

eccentric system. Indeed, the equivalence of the two systems may have been shown first by Apollonius of Perga in the third century B.C. In his *Almagest*, Ptolemy opted to represent the sun's motion by the eccentric model rather than the epicycle on a deferent, and gave as his reason the greater simplicity of using one circle rather than two, thus emphasizing the paramount Greek desire to represent natural phenomena in the simplest mathematical way.

The third model did not appear until the second century A.D. when Ptolemy proposed it in his *Almagest*. It proved to be a further refinement of the eccentric system, to which Ptolemy added what he called an "equant" point, with respect to which a planet moves with equal angles in equal times. In Figure 3.6, the equant point is located at Q. A planet moves with equal angles in equal times with respect to point Q and not with respect to points C (the center of the circle) or E (the earth).

As significant as were the advances in Greek mathematical astronomy in the Hellenistic period and later, there were other noteworthy contributions to astronomy in that same period. Apart from the further development of observational astronomy, to which Hipparchus of Nicaea (d. after 127 B.C.) made important contributions, a few Greek astronomers proposed daring new cosmological schemes. Already in the fourth century B.C., Heraclides of Pontus (c. 390–d. after 339 B.C.), a pupil of Plato, proposed, perhaps for the first time, that the earth rotates daily on its axis. Despite the fact that Heraclides' proposal failed to gain much support, Aristarchus of Samos (c. 310–230 B.C.), one of the greatest mathematicians of antiquity, proposed the first known version of the heliocentric hypothesis some 1,700 years before Nicholas Copernicus proclaimed it. Indeed, Aristarchus has often been called the Copernicus of antiquity.

Aristarchus' achievement is known to us from later writers, such as Archimedes and Plutarch (c. 100 A.D.), who tell us that Aristarchus proposed that the sun rests at the center of our cosmos and that the earth moves around the sun while rotating daily on its axis. However, the heliocentric system gained only one known supporter: Seleucus the Babylonian (c. 150 B.C.). The greatest figures of Greek astronomy—especially Hipparchus and Ptolemy—rejected it, and it was therefore never seriously considered until Copernicus proposed it as the true system of the world in his great work of 1543, *On the Revolutions of the Heavenly Spheres*.

Mechanics and Optics

The Greeks began the development of mechanics in the fourth century B.C., with contributions made by Aristotle and Archytas of Tarentum (fl. c. 375 B.C.). Indeed a treatise titled *Mechanics* has been ascribed to Aristotle, although it was more likely written by Strato of Lampsacus (d. c. 268 B.C.), one of Aristotle's successors at the Lyceum. The most significant contributors to mechanics in the Hellenistic period were Euclid and Archimedes. Euclid composed a *Treatise on the Balance* in which he proved the law of the lever geometrically, as did Archimedes in *On the Equilibrium of Planes*. In treating weights geometrically, Archimedes introduced a level of abstraction that had not been previously achieved. The same level of abstraction characterizes *On Floating Bodies*. In Proposition 7 of this work, we find the famous "principle of Archimedes," that "the solid will, when weighed in the fluid, be lighter than its weight in air by the weight of the fluid displaced" (Clagett 1957, 75).

During the third century B.C., treatises on mechanics were also written by two other significant authors, namely Ctesibius (fl. 270 B.C.) and Philo of Byzantium (fl. c. 250 B.C.). As with medicine and astronomy, the most comprehensive treatise on mechanics that has survived from the ancient world was composed after the Hellenistic period, in the first century A.D., during the period of the Roman Empire. The treatise was titled *Mechanics* and its author was Hero of Alexandria (fl. A.D. 62). Among numerous contributions, Hero describes the five simple machines—"wheel and axle, the lever, a system of pulleys, the wedge, and the screw"—and also applies the principle of the lever to the bent lever (Clagett 1957, 76). Among numerous other treatises, Hero wrote the *Pneumatics*, in which he includes discussions and experiments about vacuum, air pressure, and water pressure. Hero was not only expert at mechanics but also a fine mathematician.

Geometric optics were usually treated separately from theories of vision, but the two were nevertheless interrelated, because even in geometrical optics one had to base the geometry on some kind of vision theory. Here again, Euclid plays a major role, having written *Optics*, a treatise that is almost wholly geometrical, with little concern for theories of vision. Euclid used the law of reflection in one of his proofs, although that law had already been known to the Greeks a century or so before. As for theories of vision, Euclid largely ignored them. David Lindberg, the aknowledged authority on ancient and medieval optics, observes that "if you were willing to confine yourself to that which could be addressed geometrically, Euclid's theory was a brilliant achievement; if you were interested in any of the nongeometrical features of vision, Euclid's theory was next to useless" (Lindberg 1992, 106).

Claudius Ptolemy, who, as we saw, was the author of the greatest astronomical work of this period, also wrote the greatest optical treatise in antiquity. Thus once again, the greatest treatise in a scientific subject was composed during the Roman Empire, a period that has often been misleadingly characterized as weak in science. Ptolemy's *Optics* is not only highly mathematical but also experimental. Ptolemy used Euclid's law of reflection but went far beyond with a theory of refraction. According to Lindberg, Ptolemy "passed on to future generations a thorough understanding of the basic principles of refraction, a clear and persuasive example of experimental investigation, and an important body of quantitative data" (Lindberg 1992, 108).

GREEK SCIENCE IN THE ROMAN EMPIRE TO THE SIXTH CENTURY A.D.

The achievements of the Greeks in science during the four or five centuries prior to the advent of the Roman Empire and Christianity are truly impressive. As we saw, the Greeks laid the basis for the life sciences (medicine and biology) and the exact sciences (mathematics, astronomy, mechanics, and optics). With the emergence of the Roman Empire from the first to the sixth centuries, Greek science continued to record significant accomplishments that were on a par with the best achievements of their predecessors in the Hellenistic period. We have already mentioned Claudius Ptolemy and Galen (c. A.D. 129–c. 200), two of the greatest scientists of the ancient world. We have seen that Ptolemy composed the greatest astronomical and optical works of antiquity, and that Galen was the greatest physician and medical writer in the ancient period who left many works that shaped the subsequent history of medicine. But there were other significant contributors to the advancement of science.

In mathematics, major contributions were made by Nicomachus of Gerasa (fl. c. A.D. 100), Diophantus of Alexandria (fl. A.D. 250), Pappus of Alexandria (fl. A.D. 300–350), Theon of Alexandria (fl. second half of fourth century A.D.), Proclus (c. A.D. 410–485), Eutocius of Ascalon (b. c. A.D. 480), and Simplicius (c. A.D. 500–d. after 533). Four of these authors—Pappus, Theon, Proclus, and Simplicius—wrote commentaries on one or more books of Euclid's *Elements*. They not only commented on Euclid but also added their own mathematical thoughts and ideas. Nicomachus of Gerasa and Diophantus of Alexandria wrote treatises on arithmetic: Nicomachus' titled *Introduction to Arithmetic*, and Diophantus' treatise bearing the title *Arithmetic*. Nicomachus' *Introduction* is of little significance in its own right but was used as the basis of an important Latin work, *Arithmetic*, by Boethius in the late fifth century, which played a very important role in medieval mathematics. Diophantus extended considerably the earlier, more primitive, algebraic analysis that had been used in geometry. He introduced, apparently for the first time, algebraic notation and a sign for the unknown. Clagett informs us that "Diophantus easily solved quadratic equations (involving the unknown squared) and in one special case a cubic equation. His name is still connected with the solution of what are called indeterminate equations. There can be little doubt that Greek mathematical genius was still burning brightly at the time of Diophantus" (Clagett 1957, 117).

Other significant contributors to mathematics during the Roman Empire could be cited, but Pappus of Alexandria, who wrote in the century after Diophantus, deserves special mention. In addition to his commentaries on Euclid's *Elements* and Ptolemy's *Almagest*, Pappus' most important contribution was the *Mathematical Collection* in which he included many important ancient mathematical problems and gave solutions to those proposed by previous mathematicians. The *Collection* is, therefore, of great historical significance. But more than that, Pappus often made improvements on these proofs. Sir Thomas Heath says of Pappus' *Mathematical Collection*: "Without pretending to great originality, the whole work shows, on the part of the author, a thorough grasp of all the subjects treated, independence of judgement, mastery of technique; the style is terse and clear; in short, Pappus stands out as an accomplished and versatile mathematician, a worthy representative of the classical Greek geometry" (Heath 1921, 2:358; see also Clagett 1957, 118).

Just as Greek astronomy reached its apogee with Ptolemy's *Almagest*, Greek medicine also reached its greatest heights in the second century A.D., with the medical treatises of Galen of Pergamum (A.D. 129–d. c. 200; see Figure 3.7). Galen studied philosophy and mathematics before pursuing medicine as a career. He studied and practiced medicine in a number of cities of the ancient world, including Pergamum, Alexandria, and Rome. He wrote an incredible number of treatises, which fill more than twenty volumes in modern editions.

Galen firmly believed that philosophy is an essential part of medical education and even expounded that idea in a treatise, *That the Best Doctor is also a Philosopher*. Indeed, he is said to have composed some twenty commentaries on Aristotle's logical works. As a physician, Galen may be considered a general follower of Hippocratic medicine. He firmly believed that the basis of medicine is clinical observation and the evidence of the senses. Although other great physicians followed Galen—especially Avicenna (Ibn Sina) and Al-Razi (Rhazes, as he was known in Latin), two great Islamic physicians—it was largely Galen's ideas about anatomy and physiology that were dominant in medicine until the seventeenth century. Galen carried out vivisections on animals, especially barbary apes. In one of his best-known works, *On the Natural Faculties*, Galen includes reports of a few of his vivisections. By his day, Roman authorities forbade human dissections and vivisections, so Galen did the next best thing and used living animals. In *On the Natural Faculties*, he describes his experiment showing that in a living animal the flow of urine from the kidney to the bladder is irreversible; he later describes how he investigated the digestive processes in pigs, to whom he fed a liquid mixture of wheaten flour and water. After three or four hours, he cut them open and ex-

Figure 3.7. Galen, physician and philosopher. (Wellcome Library, London.)

amined the state of the food in their stomachs (Cohen and Drabkin 1948, 480–483). Galen did even more experimental dissection and vivisection on animals in *On Anatomical Procedures*. In that treatise, Galen also stressed the importance of knowledge of the parts of the body. "What could be more useful," he declares, "to a physician for the treatment of war-wounds, for extraction of missiles, for excision of bones ... than to know accurately all the parts of the arms and legs. .... If a man is ignorant of the position of a vital nerve, muscle, artery, or important vein, he is more likely to be responsible for the death, than for the saving, of his patients" (Lloyd 1973, 151).

Galen's approach to medicine was greatly influenced by his conviction that the world is teleological; that is, everything exists for a purpose. In a treatise titled *On the Use of Parts*, Galen explains the purpose and function of each part of the body. He seems to have believed in a creator and may have been influenced by the mystery religions of his day, which were spread across the Roman Empire, as we see from this passage in *On the Use of Parts*, where Galen describes his treatise as "a sacred book which I compose as a true hymn to him who created us: for I believe that true piety consists not in sacrificing many hecatombs of oxen to him or burning cassia and every kind of unguent, but in discovering first myself, and then showing to the rest of mankind, his wisdom, his power and his goodness" (Lloyd 1973, 151).

From the above account, it is obvious that Greek science was still a vibrant force in the Roman Empire. Greeks who lived in that empire and were interested in science and natural philosophy drew upon the achievements of their predecessors and in a number of significant instances even added to that legacy. During that same period, however—say, from Caesar Augustus in the first century A.D. to the end of the sixth century A.D.—significant changes took place that affected the status of science in Greco–Roman civilization.

Those who engaged in science and contributed to its development in one field or another during the ancient period constituted a small group with relatively little influence. Scientific subjects and ideas were undoubtedly discussed in famous philosophical schools—Plato's Academy and Aristotle's Lyceum, for example—and medical schools scattered in various cities of the Greek world. But these were not akin to modern scientific research institutions. Although some scientists and natural philosophers were in direct contact at this or that school, most contributors to science worked alone and learned their science

from manuscript copies of scientific treatises that may by good fortune have been available where they resided, or which may have been
part of their private libraries. (The quality and accuracy of those manuscripts has been discussed in chapter 1.) In sum, scientific activity in
the ancient world was a sporadic affair carried on by individuals scattered over a vast area. They sometimes sought each other out, but that
involved contending with enormous difficulties of travel and other
formidable obstacles, such as war, plague, and political upheaval. For
the most part, then, the great majority of contributors to science
worked by themselves and could only converse with colleagues if they
happened to live in the same town or city. Because books and information on any particular scientific subject were difficult to obtain, and
manuscript copies were often unreliable and uncertain, we should be
all the more impressed by the achievements of the ancients in science.
But when we reflect on ancient science, we should not think of modern laboratories and facilities, or of scientists who could focus wholly
on their research and communicate with each other almost at will. Science in the ancient world was a tenuous and ephemeral matter. Most
people were indifferent to it, and its impact was meager. It was a very
small number of Greek thinkers who laid the foundations for what
would eventually become modern science. Of that small number, a
few were especially brilliant and contributed monumentally to the advancement of science.

And so it was that the Roman Empire inherited the legacy of Greek
science. But along with the great tradition of Greek science came other
trends and tendencies from the preceding centuries to which scholars
in the Roman Empire added their own. With the exception of a few
individuals like Hero of Alexandria, Ptolemy, and Galen, there is little doubt that the creativity and originality that characterized the Hellenistic period was considerably diminished during the centuries of
the Roman Empire. There may have been a number of causes for this.
From the sixth century to the end of the first century B.C., the Greek
ideal of scientific inquiry had always been to pursue knowledge for
its own sake and thereby to increase the sum total of knowledge. This
was the ideal of Plato and Aristotle and of virtually all Greek scientists and natural philosophers. This driving force to seek knowledge
for its own sake lost its impetus during the Roman Empire. What replaced it was a desire to preserve knowledge rather than expand it.
Scholars tended to repeat what they learned from the past, especially
the words and thoughts of great figures, such as Euclid, Aristotle, and

Archimedes. They incorporated old knowledge into certain literary forms that were conducive to its preservation. To achieve this, two major genres were developed: the handbook and the commentary.

The Handbook

The handbook tradition in science and natural philosophy was aided by the fact that, already in the fourth century B.C., Greeks began writing histories of different sciences and thus summarizing the achievements of some of the great mathematicians, astronomers, mechanicians, and physicians. Eudemus of Rhodes (fl. second half of fourth century B.C.), one of Aristotle's pupils, may have been the first to write such histories. Three have been attributed to him: a *History of Arithmetic*, a *History of Geometry*, and a *History of Astronomy*. None of these works has survived, but because later writers used them, the ideas and thoughts have been preserved. Our knowledge of early geometry and astronomy derives largely from Eudemus' historical works. Such treatises were not intended as vehicles for original scientific contributions, but rather to record, and perhaps describe, those that had already been made. It was probably a relatively easy transition from historical work to compendious handbook. The purpose of a history was not only to record what had happened in a given discipline or science, but also to popularize and disseminate knowledge to those who were interested in scientific subjects but might not have been capable of understanding the technical proofs and arguments in a rigorous, original scientific treatise in mathematics or astronomy, for example.

Indeed, Aristotle, Eudemus' teacher, may have played a significant role in the handbook movement. The early histories of subjects in science and natural philosophy were of great interest to him. Gathering data on many subjects and organizing it for easy dissemination was a regular activity at Aristotle's Lyceum. Aristotle undoubtedly inspired his successor at the Lyceum, Theophrastus (c. 371–287 B.C.), who was a prolific author, especially in botany, a subject in which he was regarded as the fundamental authority until the sixteenth century. Hundreds of treatises have been attributed to him, but the one that is most relevant for the handbook tradition is his *Physical Opinions*, a work of eighteen books. However, like most ancient works, it is lost. Fortunately, it was used so frequently by later compilers that scholars have been able to reconstruct the first and eighteenth books and to obtain a good idea of the subject matter and content of the remaining books. Thus, Theophrastus' *Physical Opinions* played the role of the typical handbook of the Hellenistic period: it widely disseminated information and was used for many centuries until well into the late Roman Empire. But Theophrasus was only one source of many. There were handbooks on astronomy, arithmetic, geometry, botany, zoology, and many other topics.

Whether a handbook was written in the first century B.C. or in the fourth century A.D., it usually reflected the science of a much earlier time, often the first two centuries of the Hellenistic period. Once the handbook tradition took root, the body of information that was transmitted from one handbook author to another through the centuries tended to remain fairly constant. There were of course additions and deletions, but the main body of handbook knowledge had become rather ossified.

By the time the Romans ruled the lands around the Mediterranean Sea and came into regular contact with Greek learning, the handbook tradition was the major source of scientific and philosophical knowledge. The Romans found that format utterly congenial, as is evident from the numerous handbooks that have survived and from the titles of many that did not. They seem to have ranged over most topics (Stahl 1962, 66–67).

With the exception of those Romans who learned Greek and might have read Greek science from the original works, Roman acquaintance with Greek science was almost wholly by means of handbooks that were translated from Greek into Latin. Few treatises actually composed in Greek by Greek scientists and natural philosophers were translated into Latin. If the basic ideas of such Greek treatises became available it was by means of handbook summaries in Latin, which were usually translations from Greek handbooks, to which Latin authors might add what they wished. Although Latin handbook authors frequently cited the names of great Greek scientists, such as Eudoxus, Archimedes, Euclid, Ptolemy, and others, they did not read those treatises and probably never even saw them. It was customary to pretend that one was familiar with the treatises cited, but that was rare indeed.

At first, in the second and first centuries B.C., the Romans wrote encyclopedic treatises based on Greek handbooks. The first of these was by Cato the Elder (234–149 B.C.), who wrote on many topics but was a traditional Roman who was very hostile to Greek learning. Marcus

Terentius Varro (116–27 B.C.) was perhaps the most prolific Roman encyclopedist. He wrote a treatise titled *The Nine Books of Disciplines*, in which he presents the Greek liberal arts of rhetoric, grammar, and dialectic, which came to be known as the *trivium*, and arithmetic, geometry, astronomy, and music, which came to be known as the *quadrivium*. The trivium and quadrivium would become the basis of education in the early Middle Ages. For although Varro's *Nine Books of Disciplines* is lost, borrowings from it were so extensive that substantial portions of his legacy were preserved and proved influential in the early Middle Ages. Although others might be mentioned, Titus Lucretius Carus (c. 95–55 B.C.), or Lucretius, is one of the last of the encyclopedists of the Republican era of Rome. Lucretius is of importance because he followed the atomic theory of Epicurus in composing his famous treatise, *On the Nature of Things* (*De rerum natura*). Like Epicurus and other atomists, Lucretius believed that the universe is uncreated and contains an infinite number of worlds, a doctrine that most Christians would find objectionable. When it drew attention, Lucretius' treatise was always regarded with hostility by Christian natural philosophers. Because of its unacceptable cosmology, his work did not play any significant role in the Latin West until the Renaissance.

During the Roman Empire period, there were numerous compilations of science information. Marcus Vitruvius Pollio (first century B.C.), known as Vitruvius, wrote the famous treatise *On Architecture*; Aulus Cornelius Celsus (fl. c. A.D. 25) produced *On Medicine*, an important source of information on the history of Greek medicine; Lucius Annaeus Seneca (c. 4 B.C.–A.D. 65) wrote *Natural Questions*, a treatise on physical geography and meteorology, subjects Aristotle considered in his *Meteorology* but that Seneca never read. Like so many other Roman writers on this subject, however, Seneca may have derived much of his information from Posidonius (c. 135–c. 51 B.C.), a Greek philosopher who wrote on many topics, including geology, meteorology, and geography and who relied on Aristotle, especially in meteorology. Seneca's *Natural Questions* was quite influential during the Middle Ages. But even more influential was Gaius Plinius Secundus (c. A.D. 23–79), or Pliny, as he was known.

Pliny was probably the most unusual author in the ancient world. He wrote a lengthy treatise titled *Natural History* in thirty-seven books. William Stahl captures Pliny's prodigious energy for scholarship in these words:

Pliny used to arise at 1 or 2 A.M., sometimes at midnight, to begin his scholarly researches. If need be, he could drop off to sleep in a moment and at once resume his studies upon awakening. He began his round of official duties before dawn, making his call upon the emperor Vespasian, who also made a practice of working at night. By noon these official duties were completed and he was back at home. After a light lunch and a brief nap, or perhaps a rest in the sun, while making notes or extracts from a book that was read to him, he took a cold plunge bath, perhaps another nap or snack, and then was ready to return to his studies in earnest. No book passed through his hands without his excerpting from it; he maintained that none was so bad it did not have something of value. He worked until dinner; even during the meal a book was read to him and an amanuensis marked passages for copying. Once when a dinner guest chided the secretary for mispronouncing a word, Pliny rebuked the guest for losing ten lines by the interruption. During stays at his country villa Pliny devoted all his time to study. Only the time he was actually immersed in the bath was lost to scholarship, for while he was being rubbed down and dried a book was read to him or he dictated notes. On trips in winter his secretary was by his side in the sedan chair, wearing gloves so that the cold weather would not interfere with reading and note taking. (Stahl 1962, 102–103)

By such dedication to scholarship, Pliny produced his remarkable *Natural History*. After presenting a table of contents in the first book, Pliny devotes the second book to a summary of astronomical knowledge. The next four books are devoted to geography with the seventh concerned with man and his inventions. Books VIII to XI describe all kinds of real and imaginary animals. Botany forms the subject of Books XII–XIX, with medicines made from plants forming the theme of Books XX–XXVII. Medicines made from man and animals are considered in Books XXVIII–XXXI. Aquatic animals and their properties are treated in Book XXXII. The last five books (XXXIII–XXXVII) are concerned with metallurgy, plants, and gems (Clagett 1957, 110–111). Pliny's overall plan, which he did not rigorously follow, was to proceed from the cosmos to the earth and from the earth to the things on it, namely animals, vegetables, and minerals. In the preface, Pliny conveys something of the massive character of his treatise when he explains that he has used 100 principal authors from whom he has gathered 20,000 basic facts for his book. Modern scholars have counted 473 authors (373 may have been of secondary importance) and 34,707 facts. Whatever the number, Pliny's *Natural History* is a vast storehouse of factual information about our world, even though much of it may be unreliable, since Pliny simply copied his facts from whatever books were available to him. Pliny's influence, however, was enormous. *Natural History* was a major source of scientific information during the Middle Ages. Others who wrote Latin handbooks and encyclopedic treatises are more appropriately mentioned under the early Middle Ages.

The Commentary

During the Roman Empire and into the early Middle Ages, the overwhelming manner in which information about science was disseminated was unquestionably the handbook, or encyclopedia. But another significant form of literature also emerged during the Hellenistic period and eventually surpassed the handbook as the dominant mode of scientific dissemination. This was the commentary on a text.

In contrast to the handbook, commentaries focused on a specific text rather than presenting a wide range of information on numerous topics. The commentator's objective was to explain the meaning of the text to an audience interested in the treated author and subject. The commentator usually proceeded through the treatise section by section, explaining the text of each section and perhaps adding interpretive remarks or comparing opinions of the author with those of other past or contemporary Greek authors. In the Middle Ages, the commentary form assumed a questioning format (see chapter 6). Commentaries demanded a higher level of analytical and expository skills than handbooks. The commentator usually had some knowledge of the subject of the text and may have been a follower or student of the textual author. One of the first works to be subject to commentaries was Euclid's *Elements*. Although commentaries on that treatise were probably written in the second and first centuries B.C., those that are extant derive from the Roman Empire period, when commentaries were written by Hero of Alexandria, Pappus of Alexandria, and, in the fifth century, by Proclus, a neo-Platonic philosopher. For the information it contains about the history of geometry, Proclus' commentary is easily the most significant.

The most important commentaries for the interrelations between science and religion were those on the works of Plato and Aristotle, composed in the centuries after the birth of Christ. The most important single commentary on Plato was by Calcidius, who lived in the fourth or fifth century. Calcidius translated about two-thirds of Plato's *Timaeus* from Greek into Latin, adding a Latin commentary. Embedded in Calcidius' commentary were important descriptions of Greek astronomical knowledge in a period when such information was relatively scarce. Calcidius furnished Western Europe with its only significant direct knowledge of a Platonic treatise. His translation was important because it was the only extensive text by Plato known to readers in the Latin West for approximately 800 years.

Another significant commentator on Plato was Proclus (A.D. 410–485), a member of the Athenian school of neo-Platonists. Proclus was a prolific author, and although most of his works have been lost, we do have his commentaries on Plato's *Timaeus, Republic*, and *Parmenides*, which, however, were not known in the Latin Middle Ages.

Because Aristotle is the key figure in this study, commentaries on his works are the most relevant. During the Roman Empire, Aristotle's texts assumed great importance. Between A.D. 200 and 600, Greek philosophers and natural philosophers commented on Aristotle's texts to such an extent that approximately 15,000 pages of comments have been preserved, much of it still unknown and unanalyzed. The great production of commentary literature on Aristotle's works is largely explicable by the fact that most neo-Platonists held Aristotle's works in high regard. They were the dominant philosophical school in late antiquity. The greatest number of Aristotelian commentaries in this period came from the neo-Platonists, largely because they mistakenly regarded the thought of Plato and Aristotle as harmonious and sought to demonstrate this whenever feasible. The philosopher Porphyry (c. A.D. 232–d. c. 301–306), a student of Plotinus, the founder of neo-Platonism, was the main proponent of this tactic, which became characteristic of neo-Platonist commentators during the next few centuries. There were also genuine followers of Aristotle—usually known as Peripatetics—who added to the commentary literature. Perhaps the two most famous members of this group were Alexander of Aphrodisias (fl. second–third century A.D.), who left commentaries on Aristotle's *Metaphysics, Prior Analytics*, and *Meteorology* and a few other treatises; and Themistius (c. A.D. 317–c. 388), who wrote commentaries on Aristotle's *Physics, On the Heavens, On the Soul*, and other works. Both Alexander and Themistius influenced medieval Latin thought, largely through the Islamic natural philosopher Averroes (Ibn Rushd), who cited from both of their Aristotelian commentaries and whose works were subsequently translated from Arabic into Latin.

Although most of the neo-Platonic commentaries on Aristotle's texts that were written in the third and fourth centuries have been lost, a number from the fifth and sixth centuries have been preserved. Two commentators from this later period had a great impact on the subsequent history of the influence of Aristotle's thought in Islam and the Latin West during the Middle Ages: John Philoponus (fl. first half of sixth century A.D.) and Simplicius (c. A.D. 500–d. after 533). Philoponus is extremely important because he was both a Christian and neo-Platonic Aristotelian commentator. Unlike most Aristotelian commentators, however, Philoponus was very critical of Aristotle's ideas in physics and cosmology, some of his criticisms deriving from his Christian beliefs. Simplicius was also a neo-Platonist commentator, but he defended Aristotle against Philoponus' criticisms. Both Philoponus and Simplicius played significant roles in the history of science during the late Middle Ages. The quarrel between them is relevant to the history of science and religion and will be considered in the next chapter.

As we shall see later, the commentary form of literature played a major role during the Latin Middle Ages, when it underwent a significant transformation. Now, however, it is time to bring Christianity into the intellectual world of the Roman Empire.

Chapter 4

—•— ⊯◆⊯ —•—

The First Six Centuries of Christianity: Christian Attitudes toward Greek Philosophy and Science

Christianity emerged at a time when numerous mystery religions had also come into being. Since the Hellenistic period, Greek culture had interacted with Near Eastern cultures. When the Romans conquered the lands around the Mediterranean Sea, they ruled over disparate peoples, some of whom lived under the primary influence of Greek culture, while others, farther east, lived under the influence of ancient Near Eastern civilizations. There was an inevitable intermingling of these distinct peoples, from which new religions emerged, often known as "mystery religions." Many of these mystery religions were based on gods and goddesses from a particular region whose worship became internationalized as they were adopted by Roman soldiers and disseminated by them and other worshippers through the empire.

THE MYSTERY RELIGIONS AND ASTROLOGY

The gods and goddesses of the mystery religions could be found in many parts of the Roman Empire. Among the more popular and better known are the Great Mother of Phrygia, or Magna Mater in Latin; the Persian god Mithras, associated with the sun and open only to men (women were excluded from Mithraism); and the great cult of Isis, which transformed the Egyptian goddess into a Hellenized mystery religion in which Isis became a lunar goddess. Mystery religions were largely religions of salvation that promised their followers an immortal afterlife. They were called mystery religions because each had a ceremony that was only open to those who had been initiated into its secret rituals. "In addition to its elaborate and colorful ritual, each mystery religion had a revealed theology describing the nature and purposes of the gods and explaining the origin and subsequent history of the world and man. Each laid great emphasis upon sin and the necessity for purification, each preached various forms of asceticism, and each promised a glorious immortality to its devotees. Finally, each religion had its recognized initiates and professional clergy interested in missionary activity" (Swain 1950, 471).

One unusual mystery religion was Gnosticism, named from the Greek word *gnosis*, which means "knowledge." It was a mixture of concepts drawn from other mystery religions, astrology, and various philosophers and philosophical sects, including Plato, the Pythagoreans, and the Stoics. Included among its sacred texts was the Hermetic corpus, a collection of treatises attributed to the god Hermes Trismegistus ("thrice-great Hermes"). It included texts in alchemy, astrology, physics, botany, and medicine. In this mélange of writings were texts on magic and mysticism. By attributing all the Hermetic treatises to a god, the Gnostics hoped to lend credence to the texts and to give them an air of authority that would discourage skeptics. Most of the texts relied on supernatural causation to explain natural phenomena.

Most noteworthy about the Gnostics is the fact that they sharply distinguished between God and the world, urging believers to embrace the former and reject the latter. Thus, Gnostics rejected the physical world as evil. "They praised poverty, celibacy, and contempt of the world, and they taught a mode of life by which men might free themselves from the temptations of the 'flesh' and of 'matter' to become 'psychic' or 'spiritual'" (Swain 1950, 475). Gnostics believed that only a savior god could redeem their sins and give them immortality.

Mystery religions were common in the Roman Empire. They came to share many characteristics, especially that of rebirth after death. In the *Golden Ass* by the Roman author Apuleius, an initiate of the cult of Isis declares: "I approached the borderland of death and . . . when I had been borne through all the realms of nature I returned again; at midnight I beheld the sun blazing with bright light; I entered the presence of the gods below and the gods above and adored them face to face." Isis then promises the initiate: "When thou shalt have run the course of thy life and passed to the world beneath, there too in the very vault below the earth thou shalt see me shining amid the darkness . . . and reigning in the secret domains . . . and thyself dwelling in the fields of Elysium shalt faithfully adore me as thy protector" (Thompson and Johnson 1937, 21).

The mystery religions drastically altered the religious complexion of the ancient world. They affected both ordinary people and intellectuals. For a segment of the intellectual class, however, philosophical schools, some of which had become quasi-religious organizations, served to bring religion and philosophy together. Indeed, philosophy became a rival to Christianity. Several phases in the development of the philosophical schools can be identified in the long course of Hellenistic–Roman history. Frederick Copleston distinguishes three such phases (Copleston 1960, 382–383). The first, ranging from approximately 300 B.C. to around 50 B.C., saw the founding of the Stoic and Epicurean schools, which emphasized personal conduct and the attainment of personal happiness. From 50 B.C. to around A.D. 250, the Greco-Egyptian city of Alexandria played a significant role. There philosophical schools returned to a kind of philosophical orthodoxy by emphasizing the thoughts and ideas of their founding masters. Because some of the masters, such as Aristotle and Plato, had some interest in science, this interest was reflected in the writings of their followers. But along with this more rational interest, there was a strong trend toward religious mysticism. The neo-Pythagorean school, for example, incorporated religious mysticism into its philosophical speculations.

The third and most important phase, covering the period from around A.D. 250 to approximately A.D. 650, and the most relevant for Christianity, was the rise of the neo-Platonic school of philosophy. The founder of neo-Platonism and the neo-Platonic school of philosophy was Plotinus (c. A.D. 204–270), who was probably born in Egypt, studied philosophy at Alexandria, and eventually taught in Rome, where he settled around 243. His writings were brought together in a treatise titled *Enneads,* so titled because the six books that comprised it each contained nine tracts.

For Plotinus, God, a transcendent being called the One, is beyond all being of which we can have any experience. God is beyond all distinctions. Moreover, God does not engage in thought or in any acts of the will. No positive attributes can be assigned to the One because to do so would be to delimit and particularize God. God is absolutely unchangeable and omnipresent. And yet the world emanates from him by necessity. God does not will the world to emerge, because that would imply change in God. The things that emanate from God do not diminish him in any way whatsoever. Indeed, he is unaware of these emanations. The first emanation from God is Mind or Thought (*nous*), which is eternal and beyond time. From Mind or Thought emanates the Soul, akin to Plato's World Soul, and from Soul emanate individual souls. In Plotinus' philosophy, as in Plato's, the soul exists before its union with the body and will survive after the death of the body in a state of immortality. The material world is the last emanation and lies below the soul. Plotinus regards matter as the antithesis of the One, the last gasp of the emanation process. In the eyes of Plotinus, the material world has no positive qualities; indeed, it is utterly negative.

If philosophers reflect on the great truths of philosophy, they can prepare themselves for a mystical union with the One, which, for Plotinus, is the ultimate goal of philosophical reflection and study. Although Plotinus had essentially constructed a religion, he did not include any religious practices or rites, nor did he call for prayers or sacraments. Plotinus' philosophy was given a great boost by Porphyry of Tyre, a disciple who had studied with Plotinus for eleven years. Porphyry emphasized the religious side of philosophy, insisting that the ultimate aim of philosophy is salvation. To attain this, the soul must at all times seek what is higher and nobler, a process that requires ascetic practices and knowledge of God.

In the closing centuries of the ancient, or Greco-Roman, world, neo-Platonism was the dominant philosophy. "This final speculative effort of Ancient Philosophy," Copleston explains,

attempted to combine all the valuable elements in the philosophic and religious doctrines of East and West in one comprehensive system, practically absorbing all the philosophic Schools and dominating philosophical development for a number of centuries, so that it cannot justifiably be overlooked in a history of philosophy or be relegated to the dustbin of esoteric mysticism. Moreover, Neo-Platonism exercised a great influence on Christian speculation: we have only to think of names like those of St. Augustine and the Pseduo-Dionysius. (Copleston 1960, 383)

The many members of these mystery religions and philosophical schools had little interest in the kind of rational discourse typified traditionally by those who were engaged in science and natural philosophy. Occasionally, we find someone who was both a mystic and a rationalist, as was Iamblichus (c. A.D. 250–c. 330), a neo-Platonist born and raised in Syria. Iamblichus believed in mystery religions and magic and wrote a book titled *On the Egyptian Mysteries*. He also wrote *On the Common Mathematical Science*, in which he describes the role of mathematics in science by emphasizing mathematics as the key to understanding all natural phenomena (Lloyd 1973, 155–156).

Associated with the mystery religions, and often incorporated within them, was astrology. Mithraism, for example, was heavily infiltrated by astrological lore and doctrine (Lindsay 1971, 385–392). That the stars and planets played a role in human affairs was widely accepted in the ancient world. Astrology was an old discipline with roots in ancient Mesopotamia. In the second century A.D., Ptolemy, as we mentioned earlier, wrote not only the greatest astronomical treatise of the ancient world, the *Almagest*, but also authored the *Tetrabiblos*, or four-parted book, the greatest astrological treatise of the ancient world and still a potent force in the modern world. Although Ptolemy had a higher standard than his predecessors and contemporaries, he believed that, based on the positions of the celestial bodies, predictions of events on earth, as well as of human actions, were possible. It was Aristotle, however, who provided a theoretical basis for belief that celestial bodies influenced the terrestrial region. He believed that celestial matter is incorruptible and therefore divine. It seemed logical to him that the incorruptible celestial ether should exercise an influence over the corruptible and always changing terrestrial bodies. But Aristotle never wrote about astrology and certainly did not believe in astrological prognostication. He only believed that the celestial region influences the terrestrial region. Ptolemy not only accepted Aristotle's arguments, but went further and allowed for prediction of human events. Few astrologers, however, were as cautious and restrained as Ptolemy. Like most people who lived in the Roman Empire period, these astrologers believed in the divinity of the celestial bodies and the ability of the stars and planets to act as signs, if not the actual causes, of human actions and physical events.

THE TRIUMPH OF CHRISTIANITY IN THE ROMAN WORLD

In the century of its birth, Christianity was only one among many mystery religions that vied to recruit faithful followers and disciples. It was a time when the old traditional gods of the Greeks and Romans were being abandoned in favor of the gods and goddesses that represented the new mystery religions. Many of the beliefs that became current among the mystery religions were similar to those professed by Christians. Many regarded the world as evil and assumed that it would eventually come to an end. They further believed that a redeemer god would die so that they, his faithful followers, might enjoy eternal life. There was a widespread, and perhaps even desperate, desire to form a union with a personal god that would ultimately produce salvation and everlasting life.

From amidst the competition of these numerous mystery religions, Christianity emerged triumphant by the end of the fourth century A.D., when, in 392, the emperor Theodosius made it the state religion of the Roman Empire, while also declaring an end to pagan worship, which was thereafter regarded as treason. By contrast with Islam, Christianity was disseminated through the Roman Empire rather slowly. Within 100 years after the death of Muhammad in 632, Islam was the dominant religion over a vast area stretching from the Straits of Gibraltar to the Indus River, east of Persia. It achieved this largely by conquest. Traversing a wholly different path, and despite occasional persecutions, Christianity was spread in a relatively peaceful manner, albeit gradually. The slowness of its dissemination proved a boon to the new religion, because it allowed Christianity a lengthy period to adjust to the Roman Empire as a whole and, especially, contributed to its intellectual heritage. As a result, many educated Christians learned to live with Greek secular learning by recognizing what parts of it were actually helpful and what aspects of it to avoid or reject.

One essential aspect of Christianity that was crucial in determining its relations with the Roman government, and all subsequent governments, is the Christian insistence on the separation of church and state. This may not be surprising when one realizes that in the early centuries of Christianity, Christians were primarily interested in personal salvation and entering the kingdom of heaven. Consequently, they focused very little on relations with the state. Their emphasis lay in doing their duty to the state and hoping the state did not violate any of their doctrines and practices. Basically they hoped the state would keep out of their religious affairs and leave them to worship and organize as they saw fit.

The separation of church and state was not just a societal occurrence. It had biblical sanction. When the Pharisees asked Jesus if it was lawful to pay tribute to Caesar, Jesus replied that they "Render therefore unto Caesar the things which are Caesar's; and unto God the things that are God's" (Matt. 22:21). As a result there was, not surprisingly, conflict between the two institutions, aspects of which will be considered below. But this conflict was, in the long run, merely a significant irritant. The separation of church and state proved a boon to Western civilization. Although the church exercised great authority during the Middle Ages, and those who seemed to be subverting ecclesiastical doctrine and articles of faith were denounced as heretics and often severely punished, scholars had reasonable latitude to explore all sorts of issues, especially those that had no obvious connection with Christian doctrine. As we shall see, even where God was involved in a discussion, medieval theologians had considerable freedom to discuss a great variety of issues, many of which involved an application of natural philosophy to theology where, often enough, the discussion involved natural philosophy much more than theology.

CHRISTIANITY AND THE PAGAN INTELLECTUAL WORLD

Before describing the interaction of early Christians with pagan philosophers, it is well to remember that the sciences and philosophy were often integrated in the Greco-Roman world. In the Greek philosophical tradition, philosophy was much broader than any science or group of sciences. But most sciences, except the exact sciences of mathematics, astronomy, optics, and mechanics, were integrated within philosophy itself. In the Roman Empire period, the division of theoretical knowledge—that is, philosophy—was based largely on Aristotle's threefold division of it, as described earlier in chapter 2: metaphysics, or theology, mathematics, and physics. The third, or natural philosophy, as it came to be called, incorporated within itself parts of many sciences, such as cosmology, geology, geography, biology, chemistry, oceanography, and others that eventually became separate sciences. Natural philosophy, with all its subdivisions and parts, which sometimes included medicine and metaphysics, was the fundamental tool for investigating the physical universe, the same universe that Christians also sought to understand. Thus, when I use the term "science," I will usually mean natural philosophy as a whole, rather than any specific science. Except for the exact sciences, the term "natural philosophy" signifies all sciences indiscriminately. Indeed, during the late Middle Ages, some natural philosophers applied mathematics to physical problems and thus incorporated parts of an exact science into natural philosophy.

As Christians became more solidly established in the Roman Empire, their efforts to explain the universe and its workings based upon scripture inevitably clashed with traditional Greek interpretations. But Christians were a diverse group, and their attitudes toward, and interpretations of, the ideas and arguments embedded in the Greek tradition of science and philosophy were far from monolithic, as we shall now see.

The Attitudes of Early Christians toward Pagan Philosophy

The Greek and Latin church fathers shaped Christian attitudes toward pagan culture and learning. As one might expect, they were a disparate group with diverse backgrounds and attitudes toward the pagan philosophical legacy that varied considerably. Some were overtly hostile to pagan philosophy and urged avoidance of pagan literature. The most striking member of this group was probably Tertullian (c. A.D. 150–c. 225), who proclaimed: "What indeed has Athens to do with Jerusalem? What concord is there between the Academy and the Church? What between heretics and Christians? . . . Away with all attempts to produce a mottled Christianity of Stoic, Platonic, and dialectic composition! We want no curious disputation after possessing Christ Jesus, no inquisition after enjoying the gospel! With our faith we desire no further belief" (Tertullian n.d., 3:246). In another treatise (*Apology*, chap. 46, par. 7), Tertullian insists that "Philosophers, with mockery and contempt, out of hostility feign the truth and, in feigning it corrupt it, eager only for glory; Christians, both strive for the truth out of necessity and maintain it unspoiled, concerned for their salvation" (Sider 2001, 67). Tertullian also had harsh words for Socrates, Plato, and other Greek philosophers, and also accused them, as did other Christian writers, of borrowing many of their ideas from the Old Testament. Despite his assault on philosophy and philosophers, Tertullian was influenced by Stoic philosophy.

Some church fathers distrusted science and philosophy and sought to show its weaknesses and uncertainties, as opposed to the certainties of Christian revelation. Church fathers such as Saint Basil, Tatian, Eusebius, and Theodoret pointed to the contradictions and silliness of many scientific conclusions. Typical of such skeptical expressions was that of Saint Basil (c. A.D. 331–379), who declared that "the wise men of the Greeks wrote many works about nature, but not one account among them remained unaltered and firmly established, for the later account always overthrew the preceding one" (Basil 1963, 5).

The most productive Christian attitude, however, regarded pagan philosophy as a repository of helpful ideas for the better understanding of Christianity itself. It was philosophy in the service of Christianity, a relationship that came to be called the "handmaiden tradition." In the course of subsequent history, philosophy—by which I mean the amalgam of natural philosophy, science, and metaphysics—was destined to play a significant role in shaping Christian theology and instilling a scientific spirit and outlook in the Christian milieu, results that ultimately proved extremely beneficial to Western civilization. The basis of the Christian attitude toward philosophy was established even before Christians showed any concern for philosophy. It had Jewish roots. For "the history of Christian philosophy begins not with a Christian but with a Jew, Philo of Alexandria, elder contemporary of St. Paul" (Chadwick 1970c, 137).

Philo of Alexandria (c. 25 B.C.–c. A.D. 50), known also as Philo Judaeus, was a Hellenized Jew who knew and wrote in Greek, but probably did not know Hebrew. He was well acquainted with the philosophy of his day, especially with the works of Plato, which greatly influenced him. For Philo, philosophy was essential for understanding scripture and the higher wisdom of theology. His interest in philosophy, however, was largely guided by the kinds of questions that arose in his efforts to understand his faith.

Philo is generally regarded as the initiator of the handmaiden tradition, namely, the idea that secular disciplines, especially natural philosophy and science, should not be studied for their own sakes, but only to understand and explicate holy scripture and theology. This attitude was subsequently adopted by many church fathers.

The Autonomy of Nature

A powerful force that motivated Christians to study nature was the sense that God had created the world as an essentially self-operating entity. He had endowed it with the capacity to function by its own laws and causes. In the second century B.C., a few Jewish authors, notably Jesus ben Sirach and Aristobulus of Alexandria, held that God formed nature to operate with regularity and by its own inherent powers. This attitude is reflected a few centuries later in Saint Basil's fifth homily, when he declares that earth continually produces herbs, seeds, and trees, just as tops, after they are spun, whirl about and continue to spin. It is in this same manner that the "order of nature, having received its beginning from that first command continues to all time thereafter, until it shall reach the common consummation of all things" (Basil 1963, 82). Basil speaks of the inherent natural order on a number of other occasions (see below for his mention of the way fish migrate). We shall again meet the concept of a God-created autonomous cosmos in the writings of Saint Augustine (Kaiser 1997, 32–43). With the introduction of Aristotle's natural philosophy into the Christian West in the twelfth century, nature would become more autonomous—though not without a struggle—than any of the church fathers could have imagined.

Although God created a world that he endowed with inherent laws, which enabled it to operate in an autonomous manner, this did not mean that God would not intervene to perform miracles from time to time, as scripture clearly reveals. Thus, God responded to Joshua's request by lengthening the day and commanding the sun to stand still over Gibeon (Josh. 10:12–14), and he also promised to add fifteen years to the life of King Hezekiah and, as a sign of His intent, moved the shadow of a sundial back ten degrees (4 Kings 20:1–11). But Christian scholars understood that such deviations from the natural order were rare and special.

The Greek Church Fathers

One of the earliest, if not the earliest, church father to be guided by the hand-maiden approach to nature was Justin Martyr (d. c. A.D. 165). Justin had a positive attitude toward Greek philosophy. He believed that Socrates and Abraham were Christians before Christ. Indeed, he believed that with a few adjustments, Plato could be reconciled with Christ (Chadwick 1970a, 161, 162). Justin was convinced that the best aspects of Greek philosophy were compatible with Christianity, probably because, like many other Christians, he thought that the best of Greek philosophy was borrowed from the Old Testament. Justin thought of himself as a philosopher and regarded Christianity as the best and truest of all philosophies.

Clement of Alexandria (Titus Flavius Clemens) (c. A.D. 150–c. 219) followed in the path of Justin Martyr. He was disturbed by the fact that many in his day were skeptical about the very possibility of attaining any knowledge at all, a tendency he opposed. Indeed, he viewed Greek philosophy as an important preparation for Christianity. Like Justin Martyr, he believed that the Greek philosophers drew their ideas and inspiration from the Old Testament. Clement was also convinced that Greek philosophers had arrived at many truths because the divine *Logos* illuminated and guided them. Clement wanted to use Greek philosophy to shed light on Christian theology. In the *Miscellanies*, he sought to show the importance of philosophy for Christianity. In much of this treatise, Clement explores the relations between philosophy and Christian theology, emphasizing a cautious approach to philosophy, one in which Christians select what is useful and ignore what might prove harmful. Although Philo of Alexandria and Justin Martyr regarded Greek philosophy as a useful, if not necessary, tool for the proper understanding of their respective religions, it was Clement who provided the weightiest arguments for using philosophy to serve Christianity as the handmaid of theology. Indeed, one chapter in *Miscellanies* is titled "Philosophy the Handmaid of Theology." In that chapter, Clement shows that philosophy serves Christianity and should definitely be used wherever it might prove helpful. He opens the chapter with these words: "Accordingly, before the advent of the Lord, philosophy was necessary to the Greeks for righteousness. And now it becomes conducive to piety; being a kind of preparatory training to those who attain to faith through demonstration." Clement concludes: "philosophy, therefore, was a preparation, paving the way for him who is perfected in Christ" (Clement of Alexandria 1983, bk. 1, chap. 5:305). Clement's high opinion of philosophy is reflected in his assertion that " 'philosophy is the study of wisdom, and wisdom is the knowledge of things divine and human; and their causes.' Wisdom is therefore queen of philosophy, as philosophy is of preparatory culture" (Clement of Alexandria 1983, bk. 1, chap. 5:306).

At different places in his lengthy treatise, Clement finds occasion to praise philosophy, as when, in Book 6, he declares: "Philosophy is not, then, the product of vice, since it makes men virtuous; it follows, then, that it is the work of God, whose work it is solely to do good. And all things given by God are given and received well" (Clement 1983, bk. 1, chap. 17:517). In the last chapter of the sixth book, Clement inquires "whether we ought to philosophize" and concludes that we cannot avoid it. For "if we are not to philosophize, what then? (For no one can condemn a thing without first knowing it): the consequence, even in that case, is that we must philosophize" (Clement 1983, bk. 6, chap. 18:518).

But if Clement urged Christians to utilize Greek philosophy to properly understand the Christian faith, he also convinced himself, as did many other Christians, that Greek philosophers had drawn—nay, stolen—their knowledge from the Old Testament and the Hebrew prophets, and that most of them were not even Greeks but barbarians. The widespread belief that the Greeks had derived their philosophy from the Old Testament prompted some Christians to rationalize their use of Greek philosophy as a justifiable appropriation of something the Greeks had originally taken from the Judaic tradition. Certain biblical phrases were found convenient for sanctioning such appropriations. Christians could take what was of value in pagan thought and use it for their own benefit, just as in Exodus (3:22, 11:2, and 12:35), when the Lord instructed Moses to plunder the wealth of the Egyptians. As additional incentive, Christians were urged to use the knowledge they acquired from pagan philosophy to defeat the pagans, as David slew Goliath with Goliath's own sword (1 Sam. 17:51). Elsewhere, Clement speaks of the Greek philosophers as a group, ranging from the pre-Socratics to Plato and Aristotle, and explains that "most of them were barbarians by extraction, and were trained among barbarians" (Clement 1983, bk. 1, chap. 15:315).

Many church fathers sought to deflate Greek achievements in philosophy, either by accusing them of deriving their philosophical knowledge from the Hebrew scriptures, or by denying that they were really Greeks, viewing them instead as barbarians from various parts of the ancient world. All of this was done to make their Christian readers feel that philosophy was a useful discipline, because, after all, it had its roots in the Old Testament, and, as further comfort, they could emphasize that most of its practitioners were not even pure Greeks but were of barbarian extraction.

Writing some thirty years after Clement, Origen (c. A.D. 184–c. 254), who was probably born in Alexandria, also advocated the study of philosophy for Christians, as he indicates in a letter to the soon-to-be bishop of NeoCaesaerea, Gregory Thaumaturgus. Origen urges Gregory to bring to his bishopric

on the one hand those parts of the philosophy of the Greeks which are fit, as it were, to serve as general or preparatory studies for Christianity, and on the other hand so much of Astronomy and Geometry as may be helpful for the interpretation of the Holy Scriptures. The children of the philosophers speak of geometry and music and grammar and rhetoric and astronomy as being ancillary to philosophy; and in the same way we might speak of philosophy itself as being ancillary to Christianity. (Origen 1980, 295)

Although he emphasized the preparatory nature of Greek philosophy for the interpretation of holy scripture, Origen also emphasized some of the difficulties philosophy posed for Christianity. In a treatise titled *Against Celsus* (*Contra Celsum*), Origen explains that "philosophy and the Word of God are not always at loggerheads, neither are they always in harmony. For philosophy is neither in all things contrary to God's law nor is it in all things consonant" (Chadwick 1970b, 186). Origen proceeds to illustrate this observation by noting points of agreement and disagreement:

Many philosophers say there is one God who created the world; some have added that God both made and rules all things by his Logos. Again, in ethics and in their account of the natural world they almost all agree with us. But they disagree when they assert that matter is co-eternal with God, when they deny that providence extends below the moon, when they imagine that the power of the stars determines our lives or that the world will never come to an end. (Chadwick 1970b, 186)

A few of these disagreements, as we shall see, played a large role in the subsequent history of the relations between science and religion in late antiquity, and especially during the late Middle Ages.

The last of the Greek church fathers was John of Damascus, or John Damascene, who in the eighth century lived in areas that had been Islamic for some years. Little is known about him except that he entered the monastery of Saint Sabbas in Jerusalem around the year 730. Although John of Damascus wrote numerous works, he is best known for the *Fount of Knowledge*, a treatise written sometime after 743. The *Fount of Knowledge* is in three parts. The first is called the Philosophical Chapters and is extremely important; the second contains about 103 different heresies; and the third part, divided into four books, is titled *On the Orthodox Faith*. In the second book of *On the Orthodox Faith*, John includes a lengthy discussion of the creation and its various visible and indivisible entities. Of these three parts, the first is relevant to this section, while the creation account in the second book of the third part will be discussed below in the section on hexameral literature.

In the Philosophical Chapters, John presents a detailed account of various aspects of Greek philosophy, much of which he drew, directly or indirectly, from the works of Aristotle. In his preface, John adopts a favorable attitude toward Greek philosophy when he proclaims: "First of all I shall set forth the best contributions of the philosophers of the Greeks, because whatever there is of good has been given to men from above by God, since 'every best gift and every perfect gift is from above, coming down from the Father of lights'" (James 1:17; John of Damascus 1958, 5). In the opening words of his first chapter, John reveals a strong belief in the acquisition of knowledge and utter contempt for ignorance, declaring: "Nothing is more estimable than knowledge, for knowledge is the light of the rational soul. The opposite, which is ignorance, is darkness. Just as the absence of light is darkness, so is the absence of knowledge a darkness of the reason. Now, ignorance is proper to irrational beings, while knowledge is proper to those who are rational" (John of Damascus 1958, 7).

With this introduction to his Philosophical Chapters, John explains the aim of his efforts in the second chapter: "Our purpose, then, is to make a beginning of philosophy and to set down concisely in the present writing, as far as is possible, every sort of knowledge. For this reason let it be entitled a *Fount of Knowledge*." John then declares that his work is not based on his knowledge and ideas (indeed, he says, "I shall say nothing of my own"), "but I shall set down things which have been said in various places by wise and godly men. First of all, then, it is best to know just what philosophy is" (John of Damascus 1958, 10). In the third chapter, John describes the nature of philosophy: "Philosophy is the art of arts and the science of sciences. This is because philosophy is the principle of every art, since through it every art and science has been invented" (John of Damascus 1958, 11). He then divides philosophy into speculative and practical parts and then further subdivides each part much the way Aristotle did. And then, like Clement long before, John argues against those skeptics who argue that philosophy does not exist. "There, are, however, some people," he declares, "who have endeavored to do away entirely with philosophy by asserting that it does not exist and that neither does any knowledge or perception exist. We shall answer them by asking: How is it that you say that there is neither philosophy, nor knowledge, nor perception? Is it by your knowing and perceiving it, or is it by your not knowing and perceiving it? But if it is by your not knowing it, then no one will believe you, as long as you are discussing something of which you have no knowledge" (John of Damascus 1958, 12–13). John concludes that there is obviously such a thing as philosophy and, beginning with chapter four, launches into a discussion of that subject, starting with being, which is followed by a long series of special topics in philosophy.

Unlike the Greek church fathers who preceded him, John of Damascus expressed his views on philosophy and its utility by actually informing his readers of its substantive content. He obviously did so because he regarded philosophy as an important subject and, therefore, thought his fellow Christians should become knowledgeable about it.

The Latin Church Fathers

From around A.D. 300 to 600, a group of Christian writers known as the Latin church fathers used Latin as their language of communication and, like their counterparts among the Greek church fathers, found it necessary to consider the relations between philosophy and the doctrines of the Christian faith. They were almost all ignorant of Greek, and therefore, what knowledge they had about the Greek philosophers came to them via Latin translations from Greek, or from Roman authors like Cicero and Seneca, who transmitted some information derived from Greek philosophical sources. The names of the most eminent members of this group are: Marius Victorinus Afer (c. A.D. 280–c. 363), Saint Ambrose (c. A.D. 337–397), Saint Jerome (A.D. 340–420), Saint Augustine (A.D. 354–430), Boethius (i.e., Manlius Severinus Boethius, c. A.D. 480–524/525), Cassiodorus (c. A.D. 477–c. 570) and Gregory the Great (c. A.D. 540–604). Of these Latin church fathers, Saint Augustine exercised the greatest influence on medieval thought in the Latin West. Boethius was also highly significant, but less for his important philosophical and theological views than for his contributions to the seven liberal arts, which will be discussed in the next section. Here I shall confine my attention to Saint Augustine, whose ideas influenced Thomas Aquinas and other medieval theologians.

Although his parents were Christians, though not born into Christian families, Saint Augustine (see Figure 4.1) himself did not convert to Christianity until 386 and was not baptized until 387, when Saint Ambrose baptized him in Milan. Prior to his acceptance of Christian-

Figure 4.1.   Saint Augustine. (New Catholic Encyclopedia 2003, 1:852.)

ity, Augustine had been heavily involved in philosophy and theology, the former deriving from his involvement with neo-Platonism, the latter from his days as a member of the Manichaean religion.

Like so many Christians before him, Augustine pondered the perennial question of whether Christians should use the scientific and philosophical literature of the pagans. Was it of any use to Christians? Indeed, was it not potentially dangerous and misleading? Augustine opted for the—by then—traditional handmaiden solution to the dilemma, as is evident when he declares: "If those . . . who are called philosophers happen to have said anything that is true, and agreeable to our faith, the Platonists above all, not only should we not be afraid of them, but we should even claim back for our own use what they have said, as from its unjust possessors" (Augustine 1996, 159), which Augustine justifies by appeal to the story of the despoiling, or plundering, of the Egyptians. In conformity with his handmaiden attitude toward pagan learning, Augustine places that learning within a Christian context when he declares that "all the knowledge derived from the books of the heathen, which is indeed useful, becomes little enough if it is compared with the knowledge of the divine scriptures" (Augustine 1996, 162). Thus did Augustine advocate the use of pagan philosophy when it furthered the aims of the Christian religion, but urged avoidance of secular learning when it had no such purpose. It was surely not to be studied as an end in itself.

Augustine's sentiments about pagan learning were rather common among more learned Christians. His endorsement of the handmaiden tradition is, therefore, not in itself noteworthy. But Augustine's attitude toward other important aspects of pagan literature is important and striking. His opinion on the role of reason in matters of faith was highly influential in the subsequent history of Christianity. The function of reason in the interpretation and understanding of theology and faith was to explain vital doctrines and beliefs by revealing their truths. One could only achieve this goal, however, if one first accepted the faith as true and then applied reason and logic to its explication. For Augustine, faith must precede understanding. His conviction on this matter was but a consequence of the handmaiden approach. Reason was not to be applied to the faith by following arguments wherever they might lead. The truths to be explained were already known by revelation. Logic and reason could help explain them, but could not derive them.

Because Augustine assigned this role to reason and logic, and to rational discourse in general, he wrote favorably about logic and mathematics. He believed that "the discipline of rational discourse . . . is of the greatest value in penetrating and solving all kinds of problems which crop up in the holy literature" (Augustine 1996, 153–154). For Christians to capitalize on the rigorous inferences that could be derived from the application of logic to faith, Augustine cautions them to make certain that the propositions from which they would draw logical inferences were actually true. In an example, he assumes that someone asserts as an antecedent that "there is no resurrection of the dead," from which it follows as a consequent "that neither has Christ risen again." Augustine concedes the validity of the logic, but insists that "this consequent is false, because Christ has risen again. Therefore the antecedent is also false, that there is no resurrection of the dead; accordingly there is a resurrection of the dead. This can all be put very briefly as follows: If there is no resurrection of the dead, then neither has Christ risen again; but Christ has risen again; therefore there is a resurrection of the dead" (Augustine 1996, 155). Thus, one must begin with the true proposition that Christ has risen again and then, by the use of logic, draw the correct inference that "there is a resurrection of the dead." Augustine also exerted an important influence on the way medieval natural philosophers and theologians viewed the relationship between natural philosophy and scripture, a theme that will be considered later (see chapter 7).

Saint Augustine and the Greek church fathers we have mentioned to this point shaped the handmaiden approach to Greek secular learning. They made it respectable, and even essential, for Christian authors to study Greek philosophy and science where these were thought to contribute to the advancement of Christianity. Although these early Christian thinkers discouraged the study of Greek pagan thought for its own sake, they made a significant contribution nonetheless. In their zeal for the faith, they might have condemned all Greek and secular literature, or they might have assigned it a minor, peripheral role. Instead they gave it a central role, because they gradually came to realize that Greek metaphysics and natural philosophy were essential tools for understanding the Christian religion and the world of God's creation.

COMMENTARIES ON GENESIS (HEXAMERAL TREATISES): THE CHRISTIAN UNDERSTANDING OF THE CREATION OF THE WORLD

One of the most significant early contacts with Greek natural philosophy, and philosophy in general, came when Christians sought to explain the creation account in Genesis. They were aware of pagan explanations of the world. They knew the most basic views of Plato and Aristotle and were cognizant of how dramatically their accounts differed from the scriptural version. Christian commentaries on the six days of creation as related in Genesis were called *hexamera* (or *hexaemera*), a Greek word meaning "the six days." In these hexameral commentaries, as they were known, Christians found it essential to use what knowledge of science and natural philosophy they had absorbed from pagan sources to explain the biblical account of creation, even as they differed radically from the dominant pagan theories about the world and its origins.

The Beginnings of the Hexameral Tradition: Philo Judaeus

As with the application of the handmaiden theory to Greek philosophy, the tradition of commenting on the six days of creation in Genesis began with Philo of Alexandria (or Philo Judaeus, as he was also known), who wrote *On the Creation of the World* (*De opificio mundi*). Philo was a Hellenized Jew who wrote in Greek and knew little, if any, Hebrew. He therefore had a considerable influence on Christian writers who could read Greek, but had little influence on Jewish thought, because most Jews spoke and read Hebrew, but few read and spoke Greek. As a Hellenized Jew, Philo was familiar with the works of Plato, especially Plato's cosmological work, *Timaeus*. His commentary is a mix of the revealed religion of the Bible and Greek philosophy. Philo considered a number of problems in Genesis that had counterparts in Greek philosophy and were therefore both theological and philosophical problems that would become common discussion points in subsequent hexameral commentaries. Thus, he considered whether the world had a beginning; he asked from what was the world created: Was it from pre-existent matter or matter that God himself created? Was it created in six days or was everything created at the same time? When did time begin? Why did God create plants, herbs, and shrubs on the third day, before he created the sun and moon on the fourth day? These were a few of the questions that most commentators sought to answer.

Philo argued that because the world is visible and we perceive it, it must have had a beginning. Like Plato, Philo believed that God created the world because he is good. Much of Philo's creation account follows Plato's description of creation in the *Timaeus*. Also in conformity with Plato, Philo believed that, on the first day, God created the world from an ideal, incorporeal pattern, but unlike Plato, Philo did not believe that the pattern was eternal but rather that God created it on the first day. Philo also seems to have assumed that when God created the physical world, he did so from pre-existing matter.

Philo explained seeming inconsistencies in the creation account as

Figure 4.2. God in the act of creating the sun, moon, and stars on the fourth day of the creation account in Genesis. From a thirteenth-century *Bible moralisée* in the Bodleian Library, Oxford, England. (Bodley MSS 270b, fol. 4r.)

examples of God's power. For example, God created plants, herbs, and shrubs on the third day before he created the sun and moon on the fourth day (see Figure 4.2). Philo argues that we mortals, observing natural phenomena, "suppose that the regular movements of the heavenly bodies are the causes of all things that year by year come forth and are produced out of the earth" (Philo of Alexandria 1929, 1:35). In response to such expectations, in Philo's account God replies: "Let them . . . go back in thought to the original creation of the universe, when, before sun or moon existed, the earth bore plants of all sorts and fruits of all sorts; and having contemplated this let them form in their minds the expectation that hereafter too shall it bear these at the Father's bidding, whensoever it may please Him" (Philo of Alexandria 1929, 1:35). Christian commentators, who read Philo's commentary on Genesis, were similarly influenced by Plato. Unlike Philo, however, who had no familiarity with the works of Aristotle, a number of Church fathers revealed Aristotle's direct or indirect influence.

The Greek Church Fathers and the Hexameral Tradition

Although most authors who wrote on the creation account in Genesis, or aspects of it, did so in formal commentaries, some expressed their opinions and observations in other contexts, as is evident in the sections on John Philoponus and John of Damascus. During the Middle Ages, there were other formats for expressing opinions about the creation.

Saint Basil of Caesarea   Although Saint Basil of Caesarea, who lived in the fourth century, was not the first Greek church father to write a commentary on the six days of creation, his is the first to survive. Indeed, his work was so popular and influential that it tended to replace those hexameral treatises that had already been written. It was translated into Latin and exerted a significant influence on Saint Ambrose's hexameral treatise. Numerous Latin manuscripts of Basil's work are extant, testifying to its popularity throughout the Middle Ages.

Although we saw earlier in this chapter that Basil was scornful of Greek science and philosophy, emphasizing how they always seemed to contradict one another, he nevertheless incorporated into his hexameral treatise numerous physical concepts from Greek science. After all, Basil had been well educated in Athens and was familiar with the writings of Plato and, to some extent, perhaps even with the works of Aristotle. Basil's commentary on the six days of creation appears in the first nine homilies of a sequence that extended to twenty-two homilies. His audience consisted largely of Christian workers as well as some educated people. Basil found it necessary to interject scientific explanations for various phenomena mentioned in scripture, because his audience demanded that he do so. Following a brief explanation of the passage "Let the waters below the heavens be gathered into one place and let the dry land appear" (Gen. 1:9), Basil declares: "How much trouble you caused me in my previous lectures, demanding the reason for the invisibility of the earth, since color is naturally present in every body, and every color is perceptible to the sense of sight!" And then, just before he continues on with his analysis, Basil, with seeming apprehension, suggests, "Perhaps, my words did not seem to you to be sufficient" (Basil 1963, 56). In explaining the creation account in Genesis, it was almost unavoidable for commentators on Genesis to inject information about science and natural philosophy drawn from the only source that could provide it: the pagan Greek tradition of science and natural philosophy.

Basil informs his audience that he interprets scripture in a strict, literal sense. "When I hear 'grass,' I think of grass, and in the same manner I understand everything as it is said, a plant, a fish, a wild animal, and an ox" (Basil 1963, 135). He rejected an allegorical approach to scripture, criticizing those "who have attempted by false arguments and allegorical interpretations to bestow on the scripture a dignity of their own imagining" (Basil 1963, 136). Basil regarded topics that were not mentioned or discussed in scripture as irrelevant to Christians. For example, he mentions that many have discussed the shape of the earth—whether it is a sphere, a cylinder, or is like a flat disk. But Moses did not mention the shape of the earth and, consequently, questions about that issue are of no importance to Christians.

Because he did not wish to resort to allegorical explanations for the plants, animals, and natural phenomena mentioned in scripture, Basil could not explain them away as signifying something other than what the text actually asserted. As a consequence of his desire to treat scripture as the literal truth, Basil found it necessary to describe and explain numerous animals and plants and natural phenomena mentioned therein. In response to the command in Genesis 1:24, "Let the earth bring forth living creatures; cattle and wild beasts and crawling creatures," Basil mentions and characterizes such creatures as grasshoppers, winged insects, field mice, eels, and such animals as the ox, the ass, dogs, lions, bears, foxes, tortoises, sheep, and others. He often used these creatures to moralize about humans, as when he discoursed about the dog, declaring:

The dog is without reason but, nevertheless, he has sense reactions equivalent to reason. In fact, the dog appears to have been taught by nature what the wise of the world, who occupy themselves during life with much study, have solved with difficulty, I mean the complexities of inference. In tracking down a wild beast, if he finds the tracks separated in many directions, he traverses the paths leading each way and all but utters the syllogistic statement through his actions: 'Either the wild beast went this way,' he says, 'or this, or in that direction; but, since it is neither here nor there, it remains that he set out in that direction.' Thus, by the elimination of the false he finds the true way. What more do those do who settle down solemnly to their theories, draw lines in the dust, and then reject two of the three premises, finding the true way in the one that is left? (Basil 1963, 142–143)

Basil emphasized God's providential actions: everything has a specific purpose. God created all creatures to fulfill a specific function and endowed each of them with the proper tools to fulfill their missions. He fitted carnivorous animals with sharp teeth, but for those animals who lack sufficient teeth, such as various horned animals, "He provided with many varied receptacles for the food. Because the food is not ground sufficiently fine the first time, He has given them the power to chew again what has already been swallowed. . . . The first, second, third, and fourth stomachs in the ruminants do not remain idle, but each fulfills a necessary function." In an example that would have appealed to the nineteenth-century biologist, Jean Baptiste Lamarck (1744–1829), Basil invokes the camel's neck, which "is long in order that it may be brought to the level of his feet and he may reach the grass on which he lives" (Basil 1963, 144). In his discussion of animals, Basil may have drawn on Aristotle's biological works, especially the *History of Animals*, which contained descriptions and discussions of the habits of many animals.

Basil seems to have accepted the idea that God endowed nature with the capacity to be self-operating. Thus, he believed that all animals act according to natural law instilled in them by the creator. In the seventh homily, he speaks of migratory fish that "set out all together at one preconcerted signal. When the appointed time for breeding arrives, being roused by the common law of nature, they migrate from the different bays, hastening toward the North Sea." The fish have something to teach us. "Do not despise the fish because they are absolutely unable to speak or to reason," admonishes Basil, "but fear lest you may be even more unreasonable than they by resisting the command of the Creator. Listen to the fish, who through their actions all but utter this word: 'We set out on this long journey for the perpetuation of our species.' They do not have reason of their own, but they have the law of nature strongly established and showing what must be done" (Basil 1963, 111–112).

But how can we see anything positive and providential in the production of poisonous plants? In the fifth homily, Basil explains the biblical passage beginning with the line "Let the earth bring forth vegetation." Basil explains that "immediately with the nutritive are produced the poisonous; with the grain, the hemlock; with the other edible plants, the hellebore and leopard's bane and mandrake and poppy juice. What then? Shall we neglect to acknowledge our gratitude for the useful plants and blame our Creator for those destructive of our life?" (Basil 1963, 71). Basil repudiates such a notion and argues, "There is not one plant without worth, not one without use. Either it provides food for some animal, or it has been sought out for us by the medical profession for the relief of certain diseases. In fact, starlings eat hemlock, escaping harm from the poison because of the constitution of their bodies. . . . Hellebore is food for quails, who escape harm because of their peculiar constitution. These same plants are sometimes useful to us also. For instance, with mandrake doctors induce sleep and with opium they lull violent pains of the body. . . . So, the charge which you thought you had against the Creator has proved to be for you an additional cause for thankfulness" (Basil 1963, 72).

The most significant questions for the relations of science and religion arise in the first homily, which Basil begins with the words: "In the beginning God created the heavens and the earth" (Basil 1963, 3). To explain this momentous assertion in its many aspects, Basil found it necessary to consider many topics: whether creation was instantaneous or simultaneous, or extended over some period of time; whether the heavens were created before earth; the nature of the heavenly substance; the meaning of the firmament; what are the waters above and below the firmament; the creation of planets and stars; the location and shape of the world; the manner in which the earth is supported; clouds, vapors, and the four elements; and the creation of plants and animals, including birds and sea life. In all commentaries on Genesis, the most controversial issues were concerned with questions about the creation, which involved numerous problems about the structure of the cosmos and the beginning and end of time. These questions clashed directly with Greek science and philosophy, especially Aristotle's natural philosophy.

Without mentioning Aristotle, Basil attacks Aristotle's idea that the world had no beginning (Basil 1963, 6–7). In his first homily, Basil argues that just because the celestial bodies move around in a circle, and because a circle seemingly has no beginning, it does not follow, therefore, as Aristotle argued, that the world did not have a beginning. Although we cannot find a beginning or an end to a circle, this does not signify that it is without a beginning. "He who drew it with a center and a certain radius truly began from some point," Basil declares. "Thus, indeed, because objects, moving in a circle close in upon themselves, and the evenness of their motion is interrupted by no intervening break, do not maintain the illusion of the existence of a world without beginning and without end. 'For this world as we see it is passing away' (1 Cor. 7:31). 'Heaven and earth will pass away' (Matt. 24:35)." In this argument, Basil does not demonstrate that the world must have had a beginning, but he "demonstrates" that it does by invoking two biblical quotations and thus resolves the issue by an appeal to scriptural authority. Basil then argues that if there is a beginning of the world, there must be an end of it. Since the parts of the world are corruptible and changeable, it follows that the whole made up of those parts must also be corruptible. The world cannot be co-eternal with God, the creator of all things (Basil 1963, 7).

In the second homily, Basil insists that matter cannot be uncreated, for "if matter itself is uncreated, it is, in the first place, of equal rank with God, worthy of the same honors. What could be more impious than this, that the most extreme unsightliness . . . be considered worthy of the same superior ranking as the wise and powerful and all-good Craftsman and Creator of all things?" (Basil 1963, 23). This should be taken in conjunction with Basil's belief that a spiritual light existed in an invisible world before the creation of the visible world. This spiritual light lit the invisible world that served as a light for the "whole orderly arrangement of spiritual creatures" (Basil 1963, 9). To this, God subsequently added the visible, physical world. From these ideas, we may properly infer that Basil believed the world was created from nothing—that is, nothing material. Unlike Plato, he did not assume the existence of a pre-creation matter from which God created the world. Although Basil was not one of those church fathers who explicitly argued for a "creation from nothing," or *ex nihilo*, he nonetheless made it an implicit feature of his creation account.

In the third homily, Basil seeks to elucidate the meaning of the words "Then God said, 'Let there be a firmament in the midst of the waters to divide the waters.' And God made the firmament, dividing the waters that were below from those that were above it" (Gen. 1:6–7; Basil 1963, 42). If the firmament is spherical, however, how can the water rest on the convex surface of the heavens? Would the water not roll off? Basil dismisses this problem by interpreting the passage in a way that made physical sense but would have either amused or horrified Greek natural philosophers. "If some body appears circular to us because of an inner concavity," Basil explains,

it is not necessary for the outer surface to be made completely spherical, and the whole to be perfectly rounded and smoothly finished. Let us look, indeed, at the stone vaults of the baths and the structures of cavelike buildings which, rounded to a semicircular form according to their interior appearance, often have a flat surface on the upper sections of the roof. Therefore, let them cease making trouble for themselves or for us, alleging that water cannot be kept in the upper regions. (Basil 1963, 42)

In the course of the nine homilies devoted to his commentary on the six days of creation, Basil considered many topics and drew upon what Greek science and philosophy was available to him, probably from handbooks and from authors who passed along some of Aristotle's ideas. As Etienne Gilson has put it, Basil's nine homilies on creation "contain less the systematic exposition of a philosophy than a medley of notions related to the problems which the text of scripture happened to suggest to his mind" (Gilson 1955, 55). He rarely offers cogent, detailed arguments. In his discussions of this or that subject or theme, he was often anxious to see God's providential wisdom in all the aspects of nature. Nevertheless, although Basil subordinated science to the study of nature, he often reveals an interest in nature for its own sake. His commentary on Genesis was quite influential. One of those whom he influenced was the sixth-century Christian commentator John Philoponus, who undoubtedly produced the most unusual commentaries relevant to science and religion of any Christian in late antiquity. His ideas would prove influential among both Islamic and medieval Christian natural philosophers.

John Philoponus   Among all ancient commentators on Aristotle, John Philoponus was the most philosophical, analytical, and original. He was also a significant commentator on the six days of creation. As his first name suggests, John Philoponus (c. A.D. 490–c. 570) was probably born a Christian and seems to have been a member of the Monophysite sect, which held that Christ had only one nature, not two, a view that was declared heretical in the seventh century. His surname, Philoponus, means "lover of work" and may have been a nickname, although this is uncertain (Sorabji 1987, 5). Philoponus was a neo-Platonist philosopher who studied and taught in the neo-Platonic school of Alexandria, which by Philoponus' time had witnessed an accommodation between pagan neo-Platonists and Christian neo-Platonists to the extent that some Christians served as leaders of the school by holding the chair of philosophy, a position that Philoponus himself may have held.

In Philoponus' day, it was customary to do philosophy by writing commentaries on Plato or Aristotle. Of Philoponus' commentaries on the works of Aristotle, at least seven have survived. The most important of these for science are his commentaries on Aristotle's *Meteorology*, *On Generation and Corruption*, and the *Physics* (Books 1–4, with surviving fragments from Books 5–8). In these works, Philoponus rejected many of Aristotle's theories and replaced them with well-thought-out new theories, which exerted a significant influence on medieval natural philosophers and even influenced Galileo in the seventeenth century. Philoponus rejected Aristotle's explanation of projectile motion and replaced it with an impressed force theory that marked a significant step toward the principle of inertia. Philoponus also believed—contrary to Aristotle—that finite motion could occur in a vacuum. Finally, it is noteworthy that Philoponus, once again in opposition to Aristotle, argued that if you drop two unequal weights from the same height, they will reach the ground at approximately the same time, an experiment that Galileo is alleged to have performed from the Leaning Tower of Pisa.

Philoponus' contributions to the theme of science and religion came in a variety of works, including his Aristotelian commentaries, but his most relevant contributions derive from his commentary on the six days of creation, a work titled *On the Creation of the World* (*De opificio mundi*), and two other works, *Against Proclus, On the Eternity of the World*, written in 529, and *Against Aristotle, On the Eternity of the World* (*De Aeternitate Mundi contra Aristotelem*), which is lost, but fragments of it have been preserved in commentaries on Aristotle's *Physics* and *De caelo* by the neo-Platonic philosopher Simplicius, who was Philoponus' bitter rival. "Philoponus' philosophy of nature," it has been rightly said "was the first to combine scientific cosmology and monotheism" (Sambursky 1973, 134). In defense of Christian monotheism, Philoponus attacked the most basic features of Aristotle's cosmology and physics. Among numerous criticisms of Aristotle's cosmological opinions, two may be singled out as crucially significant: Philoponus' attacks on Aristotle's belief in the eternity of the world, and his rejection of Aristotle's profound conviction that the heavens are composed of a fifth incorruptible element, a celestial ether. The incorruptible fifth celestial element stands in stark contrast to the four terrestrial elements—earth, water, air, and fire—which form the matter of all changeable compounds in the region below the concave surface of the lunar sphere.

In order to support the Christian belief in a creation, Philoponus knew that he had to show the implausible, and even impossible, features of Aristotle's arguments in favor of an uncreated, eternal universe. Aristotle had argued that the world could not have come from some prior state of material existence, say A, because we would then have to inquire from whence did A come? If we say that A came from a previously existent matter, B, we then have to seek the generator, or cause, of B, and so on, leading to an infinite regress of causes, and it would be impossible to reach the beginning of the first material existence. For other reasons as well, Aristotle concluded that the physical world could not have had a beginning. The option of creation from nothing (*ex nihilo*) would have made no sense to him, and he makes no mention of it. Moreover, Aristotle had also distinguished two different kinds of infinites. The first is an *actual infinite*, an infinite that embraces everything, leaving nothing outside. Aristotle denied the existence of an actual infinite. The second infinite is a *potential infinite*, which does exist and which Aristotle describes as follows (*Physics* 3.6.206a.27–29): "The infinite has this mode of existence: one thing is always being taken after another, and each thing that is taken is always finite, but always different" (Aristotle 1984, 351). Richard Sorabji explains that Aristotle's potential infinite always has something outside of it, for "however large a finite number you have taken, you can take more. It is the reference to the possibility of taking more that guarantees this infinity will always have something outside it" (Sorabji 1987, 168).

Aristotle's belief in the eternity of the world was based upon his conviction that a potential infinite always has something outside of it. It was this belief that Philoponus subverts in *Against Proclus, On the Eternity of the World*. Philoponus first argues that if the cosmos is uncreated and has no beginning, an actual infinity of years must have passed, and therefore, an infinite number of individuals will have come into being. "But," he declares, "it is in no way possible for the infinite to exist in actuality, neither by existing all at once, nor by coming into being part at a time, as we shall show more completely, God willing, in what follows" (Sorabji 1983, 214). Philoponus shows this by arguing that even if the infinite

comes into being part at a time, one unit always existing after another, so that eventually an actual infinity of units will have come into being, then even if it does not exist all together at once (since some units will have ceased when others exist), none the less it will have come to be traversed. And that is impossible: traversing the infinite and, so to speak, counting it off unit by unit, even if the one who does the counting is everlasting. For by nature, the infinite cannot be traversed, or it would not be infinite. (Sorabji 1983, 215)

Philoponus then applies this reasoning to the human race. For if the infinite is not traversable, "but the succession of the race has proceeded one individual at a time, and come down through an infinity of individuals to those who exist now, then the infinite has come to be traversed, which is impossible. So the number of earlier individuals is not infinite. If it were, the succession of the race would not have come down as far as each of us, since it is impossible to traverse the infinite" (Sorabji 1983, 215). Moreover, if the cosmos had no beginning, and all the individuals who lived to the time of Socrates comprised an infinite number, then if all those who were born after Socrates were added to the infinite number of humans who lived to the time of Socrates, we would have a total number of individuals that exceeds the infinite; therefore, something greater than an infinite would exist, which is impossible.

The problems quickly compound. Philoponus adds that humans would not be the only infinite entity. What about the infinity of horses that would have lived if the world were uncreated and eternal? They would comprise a second infinite number of beings. The same may be said for dogs, which would constitute a third kind of infinite being, and so on. These are further impossible and absurd consequences of an eternal, uncreated world.

A final, major inconsistency with the concept of an eternal world involves the periodic revolutions of the planets around the earth at the center, which differ from each other considerably. Saturn takes thirty years to complete a revolution; Jupiter, twelve years; the sun, one year; and so on. If the heavenly motions did not have a beginning, Saturn would have made an infinite number of revolutions. If so, then Jupiter would have made almost 3 times as many revolutions as Saturn, and the sun would have made 30 times as many, and the moon, 360 times as many, and finally, the fixed stars would have made 10,000 as many as Saturn. "Is this not beyond all absurdity," Philoponus declares, "if the infinite cannot be traversed even once, to entertain ten thousand times infinity, or rather infinity times infinity," from which he concludes that "the revolution of the heavens should have had a beginning of its existence without having existed previously" (Sorabji 1983, 215–216).

Philoponus' monumental contribution to the theory of eternity or the creation of the world has been justly praised by Richard Sorabji, who explains that Philoponus "found a contradiction at the heart of paganism, a contradiction between their concept of infinity and their denial of a beginning. This contradiction had gone unnoticed for 850 years. . . . For the first time, he put Christianity on the offensive in the debate on whether the universe has a beginning. This might well be called a turning point in the history of philosophy" (Sorabji 1987, 177–178).

Although the issue of the eternity of the world versus its creation was undoubtedly the most important issue for Philoponus, and for Christians generally, Philoponus also found that a major feature of Aristotle's cosmos posed a disturbing and challenging problem, one related to the eternity of the world. He rejected Aristotle's division of the cosmos into two radically separate regions, terrestrial and celestial, the former comprised of corruptible, material bodies made up of various mixtures of the four elements, the latter composed of a fifth element, called an ether, which was material but incorruptible and weightless, and which extended from the concave surface of the lunar sphere to the outermost sphere of the fixed stars. Aristotle regarded the celestial fifth element as divine. If Philoponus wished to defend his arguments against eternity and in favor of creation, it was essential that he show that the heavens were not incorruptible and ungenerated.

To achieve this, Philoponus contended that the celestial region is not immune to change. He argued that the planets differ in color and therefore must also differ in composition. Hence there cannot be a uniform fifth celestial element. In fact, argued Philoponus, the heavens are composed of the same four elements that comprise the bodies of the terrestrial region. The four elements that make up the celestial bodies are the purest of their respective species, but they are nonetheless the same four elements. Of these four elements, however, fire predominates. For the most part, Philoponus resorted to an explanation similar to Plato's. He repudiated Aristotle on one of the most basic features of the latter's cosmology. The world is not divided into radically different incommensurable terrestrial and celestial parts, but is rather a single, unified whole.

Philoponus' arguments against Aristotle's doctrine of the eternity of the world were also relevant to the doctrine of "creation from nothing" (*ex nihilo*). Until 529, when Philoponus wrote his attack on Proclus, a pagan neo-Platonic philosopher, Christian authors were unsure of themselves about the problem of creation. They were certain that God had created the world and that therefore the world obviously had a beginning. Many pagans also believed our present world had a beginning. But they were convinced that whoever made the world made it from a pre-existing matter that was eternal and, therefore, had no beginning. In discussing the Creation, many early Christians were influenced by Plato, who argued for a pre-existent matter that the creator God, or Demiurge, fashioned into a world. Some Christian philosophers did opt for a creation from nothing. As early as the second century, Theophilus of Antioch (fl. 181) wrote an apologetic treatise titled *To Autolycus* in which he argued for a creation from nothing. But there was always ambivalence and uncertainty, as we saw earlier, when Basil seemed to accept a creation from nothing, but did not make it explicit and was perhaps unaware that he had arrived at that conclusion. Indeed, there were even biblical passages that suggested our world was created from a pre-existing chaos (Job 28, 38; Wisd. of Sol. 11:17; see Sorabji 1983, 194). Despite these passages, the creation *ex nihilo* doctrine would probably have triumphed because it had a virtually irresistible appeal on the straightforward grounds that a deity who could create a world from nothing would appear far more powerful than a deity who could only create a world from pre-existing matter.

It was, however, John Philoponus who provided the reasoned arguments that made creation *ex nihilo* a more feasible alternative. To achieve this, he had to show that the conception of God creating the world from a pre-existent chaotic matter, much as Plato had assumed, was implausible. He did this by mustering a series of powerful arguments against Aristotle's idea of an uncreated, eternally existent world. We know from his pagan Greek rival, Simplicius, that Philoponus believed in a creation from nothing. He denied that God had created the world from pre-existing matter and argued instead that God had created the matter and form of all things from nothing (Sambursky 1973, 135).

By virtue of his arguments against an infinite regress of material causes, and his arguments that revealed grave inconsistencies in the concept of an uncreated, eternal world, Philoponus prepared the way for a defense of his own belief in creation from nothing and made it more acceptable to those who had previously lacked sufficiently powerful arguments to reject an uncreated world.

Saint John of Damascus   John of Damascus devoted chapters five to ten of the second book of his *Orthodox Faith* to the creation of the physical world. This was a small portion of the whole. In numerous places John presents various alternatives without explicitly choosing from among them. In describing the nature of the firmament, John declares: "Instructed by sacred Scripture, the divine Basil says that its substance is subtile—like smoke, as it were. Others say that it is watery because it was made in the midst of the waters. And others say that it is made from the four elements. Still others say that it is a fifth body and distinct from the four elements" (John of Damascus 1958, 211). In this instance, John probably sided with Basil, largely because he informs his readers that Basil was "instructed by sacred Scripture," something he does not say about the authors of the other opinions. We observe also that he mentions the two basic rival theories of Plato and Aristotle—the former defending the four element theory and the latter proposing a fifth element.

John reports that "some have surmised that the heavens surround the universe and have the form of a sphere which is everywhere the highest point, while the center of the space enclosed by it is the lowest point" (John of Damascus 1958, 211). John goes on to elaborate that "the heavens have seven spheres, one above the other," presumably a reference to the seven planetary spheres (Moon, Mercury, Venus, Sun, Mars, Jupiter, and Saturn), excluding the outermost sphere of the fixed stars. Following Basil, John declares: "The substance of the heavens is very subtile, like smoke, and that in each one of the spheres is a planet" (John of Damascus 1958, 212). But then, John mentions another interpretation, one in which "others . . . have imagined the heavens to have the form of a hemisphere, because the inspired David says: 'Who stretchest out the heaven like a pavilion,' [Ps. 103:2] which means a tent; and the blessed Isaias: 'He that establisheth the heavens like a vault' [Isa. 40:22 (Septuagint)]" (John of Damascus 1958, 212–213). John does not choose between the two interpretations, resting content to assert that "whichever way it may be, all things have been made and established by the command of God and have their foundation in the divine will and desire" (John of Damascus 1958, 213). Most Christian commentators opted for spherical heavens, which were an integral feature of Greek cosmology and astronomy.

John considers whether certain biblical passages that seem to attribute animation to the celestial bodies are to be taken literally. Indeed, he also cites passages in which the earth and the sea are made to seem animate, for example "Let the heavens rejoice, and let the earth be glad" (Ps. 95:11) and "The sea saw and fled" (Ps. 113:3). These lines and many that are similar are not to be taken literally because, as John puts it, "Scripture can personify inanimate things and talk about them as if they were alive" (John of Damascus 1958, 214).

In the remaining chapters concerned with creation (bk. 2, chap. 7

and 10), John devotes one chapter to each of the four elements: fire, air, water, and earth, in that order. Under fire, John discusses light and the seven planets, with special consideration given to the sun and moon, comets, and astrological causation. "We say," declares John, "that the stars do not cause anything to happen, whether it be the production of things that are made, or events, or the destruction of things that are destroyed." But John allows that the planets can serve as signs for war, but then concludes, "Nevertheless, habits are something under our own control, for, in so far as they are subject to the reason, they may be controlled and cultivated by it" (John of Damascus 1958, 219).

In the *Orthodox Faith*, John of Damascus did not have much to say about the creation but apparently wished to convey some knowledge about the physical aspects of that divine act to his readers in a treatise that was not a commentary on the six days of creation.

The Latin Church Fathers on the Creation

The two most important Latin church fathers to write hexameral treatises were Saint Ambrose and Saint Augustine. Ambrose completed his around 387, the very year he baptized Saint Augustine into the Christian Church in the city of Milan.

St. Ambrose   Like most church fathers, Ambrose did not write an integrated philosophical and scientific analysis of the creation, but discussed a whole series of topics that he felt were important to his audience. In the first homily concerned with the first day, Ambrose mentions different interpretations by philosophers about the status of the world. Some, like Aristotle, said the world is eternal; others, like Plato, said the world did not always exist in the past but will always exist in the future; Democritus said there are innumerable worlds (Ambrose 1961, 3–4). Ambrose presents the standard Christian response. Moses declared, "In the beginning God created heaven and earth," and then Moses "linked together the beginnings of things, the Creator of the world, and the creation of matter in order that you might understand that God existed before the beginning of the world or that He was himself the beginning of all things" (Ambrose 1961, 5). Ambrose makes no mention of a creation from nothing, but that may be implied by the linkage of the Creator of the world and the creation of matter.

It is very unlikely that Ambrose would have understood the "creation of matter" as a creation from a pre-existent matter.

Following Basil, Ambrose regarded the heavens to be like smoke, and therefore not solid but constituted of a subtle matter. Also like Basil, Ambrose regarded topics that the Bible ignores as irrelevant. "On the nature and position of the earth," he declares, "there should be no need to enter into discussion at this point with respect to what is to come. It is sufficient for our information to state what the text of the Holy Scriptures establishes, namely, that 'he hangeth the earth upon nothing' [Job 26:7]" (Ambrose 1961, 20). After mentioning a few theories about how the earth remains at rest in the middle of the world, Ambrose remarks, "The earth is therefore not suspended in the middle of the universe like a balance hung in equilibrium, but the majesty of God holds it together by the law of His own will, so that what is steadfast should prevail over the void and the unstable. The Prophet David also bears witness to this when he says: 'He has founded the earth upon its own bases: it shall not be moved for ever and ever.' [Ps. 103:5]" (Ambrose 1961, 21). In this instance, Ambrose does not treat the earth as if it were endowed with powers to maintain itself at the center of the world, but he simply assumes that "by the will of God . . . the earth remains immovable" (Ambrose 1961, 22). As did most Christian commentators, Ambrose rejects Aristotle's incorruptible celestial ether, because this would imply the incorruptibility of the heavens. But the world is corruptible as the Lord said: "Heaven and earth will pass away, but my words will not pass away [Matt. 24:35]" (Ambrose 1961, 25).

Ambrose also rejected the famous Pythagorean claim that the celestial spheres made music as they performed their revolutions, a music that we do not hear because "we have become accustomed to that sound from the first moment of our birth" (Ambrose 1961, 50). He finds this a ridiculous claim, for "experience itself presents us an easy rebuttal to their arguments. We are able to hear thunderbolts produced by the collision of clouds; how, then, are we unable to hear the revolution of such mighty spheres which, in proportion surely to their swifter motion, should produce sounds all the more resounding?" He concludes that it is better to ignore such subjects; they "should be left to those 'who are outside.' We should adhere closely to the doctrine laid down by the celestial Scriptures" (Ambrose 1961, 51).

Accepting Basil's argument about the waters above the firmament,

Ambrose insists that the waters can indeed remain on the convex surface of a spherical celestial sphere, because "there are a great many buildings which are round in the exterior but are square-shaped within, and vice-versa. These buildings have high level places on top, where water usually collects" (Ambrose 1961, 53).

After citing biblical passages that speak of the sun, moon, and stars as signs, Ambrose considers the legitimacy of astrological predictions and forecasting. He cites Genesis 1:14: "Let them serve as signs and for the fixing of seasons, days and years," and Luke 21:25, where the Lord says, "And there will be signs in the sun, moon, and stars." But Ambrose rejects astrological predictions at the outset when he declares

some men have attempted to set down the characteristics of birth days and the future state of each newborn child. Yet a prognostication of this sort is both vain and useless to those who seek it and is an impossibility for those who promise it. What is so inane as to suppose that everyone should be convinced that he is what his birth has made him? No one, then, ought to change his condition of life and his habits or strive to become better, but rather remain in that conviction. (Ambrose 1961, 135)

In keeping with the generally hostile attitude of the church fathers toward astrology, Ambrose was also an opponent of that ancient discipline, viewing it as a pagan study that falsely attributed independent powers to the individual planets and stars, as well as to the entire celestial region.

Saint Augustine   Saint Augustine discussed ideas relevant to the creation in a number of treatises, but did so most extensively in his commentary on Genesis titled *The Literal Meaning of Genesis*, which he wrote sometime around 391 but left unfinished. In keeping with the Christian tradition, Augustine holds that God created the world from nothing. "But we must not suppose," he explains, "that unformed matter is prior in time to things that are formed; both the thing made and the matter from which it was made were created together" (Augustine 1982, 1:36). Augustine reiterates this point in the next paragraph, declaring, "God created together both the matter which He formed and the objects into which He formed it." Thus, God did not first create a formless matter, and then add the forms to shape that matter into objects, but at the outset created objects from nothing, complete with their matter and form (see Figure 4.3).

Augustine confronts an apparent inconsistency in the creation account. Genesis speaks of a creation spread over six days, but in the book of Ecclesiasticus (18:1) we are told, "He that lives forever created all things together," that is, simultaneously and even instantaneously. To reconcile these seemingly conflicting alternatives, Augustine explains that God did indeed create all things simultaneously, but chose to narrate the creation on a day-by-day basis, because "those who cannot understand the meaning of the text, *He created all things together*, cannot arrive at the meaning of Scripture unless the narrative proceeds slowly step by step" (Augustine 1982, 1:142). Pursuing this further, Augustine titles the next chapter (bk. 4, chap. 34), "All things were made both simultaneously and in six days" (Augustine 1982, 1:143; Grant 1994, 84–85).

God made the world all at once, but he also made a "before" and "after." Although all things were created simultaneously, they were also created in the order described in Genesis. Thus, the world was not only created in an instant, but that instant included a "before" and "after" of all the events described in the six days of creation. The things created over the six days were all simultaneously created in the first instant of creation. They were all embodied in what Augustine called the "seminal reasons" (*rationes seminales*); that is, God created all things in seeds that would develop over the course of time in ways that God had specified. The six days of creation were, therefore, an unfolding of these seeds into the great variety of beings that fill the earth. God created all things in the order described in Genesis, but he did so in an instant so that before and after were indistinguishable. In this account, Augustine reveals a firm belief that God created nature to function in a lawful manner. "Each species, then, with all its future developments and particular members, was created at the beginning in the appropriate seminal reason" (Copleston 1957, 77). Thus, God conferred upon nature a capacity for continuous development and in this sense seems to have given nature various powers for self-development. Augustine's conception of a simultaneous creation of all things was probably the most widely held interpretation of the creation during the Middle Ages.

Augustine considered a number of topics about the physical universe that were relevant to scripture. He accepted the spherical shape of the cosmos, despite the fact that, as we saw earlier, two biblical pas-

Figure 4.3. Using a compass, God designs the universe. From Österreichische Nationalbibliothek (Austrian National Library), Vienna, Latin MSS, MS. 2554, fol. 1r.

sages gave prima facie reasons to believe otherwise. In the one, the heaven is likened to a stretched skin (Ps. 103:2), and in the other (Isa. 40:22), it is described as shaped like a vault. Not only do these two interpretations have to be reconciled, but "it is also necessary that both of these passages should not contradict the theories that may be supported by true evidence, by which heaven is said to be curved on all sides in the shape of a sphere, provided only that this is proved" (Augustine 1982, 1:59). To counter literal-minded biblical interpreters, Augustine explains how one can reconcile the literal meanings of these two passages with a spherical world. "If a vault can be not only curved but also flat," he declares, "a skin surely can be stretched out not only on a flat plane but also in a spherical shape. Thus, for instance, a leather bottle and an inflated ball are both made of skin" (Augustine 1982, 1:60).

Topics that were not directly pertinent to salvation were regarded as irrelevant and not worth discussing. In this category, Augustine includes the motion of the heaven, or firmament. If it is stationary, how do the stars and planets, thought to be embedded in it, travel from east to west? "My reply," Augustine responds, "is that there is a great deal of subtle and learned enquiry into these questions for the purpose of arriving at a true view of the matter; but I have no further time to go into these questions and discuss them, nor should they have time whom I wish to see instructed for their own salvation and for what is necessary and useful in the Church" (Augustine 1982, 1:60–61). Augustine explains that the scholars who have studied the matter "have concluded that if the stars alone were moved while the heavens were motionless, all the known phenomena observed in the motions of the stars might have taken place" (Augustine 1982, 1:61).

Another theme that Augustine ignored concerned the shape of the earth. Whereas Basil mentioned the shape of the earth as a topic that Moses did not discuss and concluded that it was therefore not worth exploring, Augustine fails even to mention this important topic.

Augustine was fascinated by the concept of time and wrote much about it, especially about its psychological nature. In his commentary on Genesis, he informs his readers that time began with the creation of the world. It began with the movement of the creatures God created. "It is idle to look for time before creation, as if time can be found before time" (Augustine 1982, 1:153). In his famous *Confessions*, Book XI, Augustine attempts to characterize our understanding of time, explaining how we perceive its elusive and mysterious nature with respect to past, present, and future. In his commentary on the creation account in Genesis, Augustine covered numerous topics on the physical aspects of creation, though without the detail that Saint Basil supplied.

We can do no better in concluding this chapter than to cite D.S. Wallace-Hadrill's judgment on the Greek church fathers, which applies to the Latin fathers as well and which elegantly reiterates the major theme of this chapter, namely that natural philosophy and science should not be studied for their own sakes but only to acquire insight into, and knowledge about, holy scripture. In short, they must serve only as the handmaidens of theology. "The Greek fathers," Wallace-Hadrill begins,

for all their intense appreciation of nature, for all their interest in the structure and processes of nature and their insistence upon nature as a means by which God reveals his nature, nevertheless hold that god and nature are not identical, and that the mind must penetrate nature to find God. The beautiful, the useful, the intellectually fascinating, even the spiritually beneficial— all these characteristics of nature can, if allowed to become an end in themselves, distract the mind from its proper activity, the knowledge of God. Nature must not be permitted to make too great demands which might impede the forward movement of understanding. From this arises the tension, which is characteristic of Christianity from the New Testament onwards, between deep appreciation of nature on the one hand and a refusal on the other hand to be side-tracked or delayed by its beauty. The fathers follow the New Testament closely in exhibiting a disturbing oscillation between world acceptance and world renunciation. (Wallace-Hadrill 1968, 129–130)

The attitudes espoused by the church fathers in the first six centuries of Christianity would undergo a significant change in the Middle Ages. The first signs of that change appear in the early Middle Ages and come to maturity in the late Middle Ages. The interaction and intermingling of church doctrine and tradition with secular learning lies at the heart of the significant changes that would alter Western civilization. We must now describe this process.

Chapter 5

⊷ ⊶⊱⊰⊷ ⊶

The Emergence of a New Europe after the Barbarian Invasions

Until the late twelfth century, knowledge of science and natural philosophy in the Latin West was at a low level. Very little of Greek science had been translated into Latin. What was available in the Latin language was largely drawn from handbooks and compendia that had been available since the Hellenistic period and which had undergone their own evolution in the Latin-speaking world. Up to this point, we have described the level of Greek science in late antiquity and the science and natural philosophy embedded in the commentary literature on the creation account in Genesis, the so-called hexameral literature.

THE LATIN ENCYCLOPEDISTS

It will now be useful to describe, briefly, the Latin literature that served as science and natural philosophy in the period between the first and twelfth centuries. Indeed, we have already mentioned the first-century Roman authors Seneca, Pliny, and Celsus, who were not Christians and whose works were available throughout the early and late Middle Ages. But a number of encyclopedic authors who wrote between the fourth and eighth centuries provided much of the knowledge about the physical world that sustained learning until the twelfth century. The most significant were Calcidius (fl. fourth or fifth century A.D.), Macrobius (fl. A.D. 400), Martianus Capella (fl. c. A.D. 365–440), Boethius (c. A.D. 480–525), Cassiodorus (c. A.D. 480–c. 575), Isidore of Seville (c. A.D. 560–636), and the Venerable Bede (A.D. 672–735). Of this group, the last four were definitely Christians. Among the first three, Calcidius and Macrobius may have been Christians, although this has never been confirmed. No good arguments have been advanced that would lead us to believe that Martianus Capella was a Christian, and he is customarily regarded as a pagan.

Calcidius, Macrobius, and Martianus Capella

Whether Christians or not, Calcidius, Macrobius, and Martianus Capella were neo-Platonists who wrote important encyclopedic treatises in Latin that served to transmit Plato's cosmology to the Latin-speaking world. Calcidius made the most significant contribution by translating the first two-thirds of Plato's *Timaeus* from Greek into Latin and then adding his own commentary, which was approximately six times as long as Plato's text. In this way, Calcidius passed on information about Plato's views on the creation of the world and the World Soul, and also included much information about astronomy. His translation and commentary were popular, and most medieval libraries possessed at least one copy (Stahl 1971a, 14–15).

The only work relevant for science by Ambrosius Theodosius Macrobius is his *Commentary on the Dream of Scipio*, which is actually Book VI of Cicero's *Republic*. Next to Calcidius' *Commentary* on Plato's *Timaeus*, William Stahl regards Macrobius' *Commentary* as "the most important source of Platonism in the Latin West in the Middle Ages" (Macrobius 1952, 10). Like so many commentators, Macrobius' commentary is many times longer than Cicero's text—approximately sixteen times. Macrobius is important because nearly half of his commentary is devoted to cosmology and astronomy. Macrobius, as did the ancient Greeks and almost everybody in his day, regarded the earth as a sphere located in the center of the world encircled by the seven planets each in its own sphere. Much astronomical information—much of it mistaken—is incorporated into Macrobius' commentary. Although it is not likely that Macrobius read Plato or Aristotle, he was a firm supporter of Plato and a severe critic of Aristotle. Macrobius derived most of his information about Plato from Porphyry's *Commentary on Plato's "Timaeus"* (Macrobius 1952, 36). Macrobius' *Commentary* was enormously popular because it ranged over so many topics and provided information in an age when information was difficult to obtain.

Martianus Capella was one of the key figures in the preservation of science in the early Middle Ages. As is the case with most of these early authors, very little is known for certain about Martianus Capella's life. He was probably a rhetorician who, sometime between 410 and 439, wrote an enormously influential treatise titled *The Marriage of Philology and Mercury*. The *Marriage* is an allegorical treatise describing the marriage of Philology and Mercury in heaven. The wedding involves seven bridesmaids, each of whom represents one of the seven liberal arts: grammar, rhetoric, and dialectic (or logic), which were known collectively as the trivium; and arithmetic, music, astronomy, and geometry, which were grouped together as the quadrivium. In the quadrivium, Martianus transmitted a great deal of scientific information. He seems to have written his treatise to serve as a textbook for schools. As with most authors who wrote on science and natural philosophy in the fourth to eighth centuries, Martianus derived his information from handbooks and certainly did not read the books of the famous authors he mentions. Although he wrote on geometry, Martianus did not know Euclid's *Elements*; although he wrote on astronomy, he certainly did not know Ptolemy's works or the works of any other legitimate astronomer; and the same can be said for arithmetic and music.

Martianus' book has puzzled modern scholars. They are "at a loss to explain how a book so dull and difficult could have been one of the most popular books of Western Europe for nearly a thousand years." But surprisingly, medieval students seeking an introduction to the liberal arts found in "Martianus' work a fairly compact treatise dressed in fantasy and allegory" and "were both charmed and edified by it" (Stahl 1971b, 21). In a time when books of any kind were scarce and difficult to obtain, Martianus' *Marriage* served an admirable and essential service. He provided his readers with a descriptive account of the major subjects deemed essential for a proper education. The quadrivium of the seven liberal arts provided as much science as most students could have absorbed. Of the four sciences Martianus covered, scholars are generally agreed that his treatment of astronomy was the most learned and reliable. In it, he considers most of the conventional topics: "the celestial circles; northern and southern constellations; hours of daylight at the various latitudes; anomalies of the four seasons; and a discussion of the orbits of each of the planets, including the sun and moon" (Stahl 1974, 141). Indeed, Martianus declared that Venus and Mercury had heliocentric rather than geocentric orbits, a view that drew praise from Nicholas Copernicus in *On the Revolutions of the Heavenly Orbs* (1543). It was Copernicus who abandoned the universally accepted geocentric system held by Aristotle and Ptolemy and replaced it with the now accepted sun-centered, or heliocentric, system.

Boethius

Descended from an illustrious Roman family, Boethius wrote treatises relevant to both the seven liberal arts and theology and was very influential in both areas. Unlike most of his Christian contemporaries and predecessors, Boethius knew Greek and used that knowledge to translate five of Aristotle's logical works, namely *Categories*, *On Interpretation*, *Sophistical Refutations*, *Prior Analytics*, and *Topics*. (He may also have translated Euclid's *Elements* and perhaps also a few works by Archimedes.) Boethius not only commented on four logical works but also wrote five independent treatises on logic. When taken together, all of these works and a few others were known as the "old logic" (*logica vetus*), which, with its emphasis on rationality, served as a major source of intellectual activity during the darkest days of Western civilization, between the sixth to eleventh centuries. The old logic played a fundamental role in paving the way for a new emphasis on reason and reasoned argument in the eleventh century.

Boethius also wrote treatises on the quadrivum of the seven liberal arts. Still extant are *On Arithmetic* and *On Music*, which were influential throughout the Middle Ages. The former was basically an adaptation of a Greek treatise by Nicomachus of Gerasa (fl. c. A.D. 100), titled *Introduction to Arithmetic*. If Boethius wrote works on the other two disciplines of the *quadrivium*, astronomy and geometry, they have not survived.

Boethius' penchant for logic spilled over into his five theological tractates. Boethius insisted on applying logic and reason to theological problems. In *On the Trinity*, he applied reason and logic to the doctrine of the Trinity, which had produced numerous heresies largely because it was a central doctrine of Western Christianity. Boethius was convinced that all inquiries should be pursued "only so far as the insight of man's reason is allowed to climb the height of heavenly knowledge" (Boethius 1973, 5). Reason should be applied to all investigations, even if the investigation fails. Boethius began a trend that would find imitators in the long theological tradition of the Middle Ages: He applied some of the structural elements of Euclid's geometry to theology. In his third theological tractate, known as *Quomodo substantiae*, Boethius enunciated nine axioms and definitions from which he drew a theological conclusion, just as a geometer might have done in a geometrical theorem. Like Saint Augustine before him, Boethius believed that the application of reason to theology was essential. Henry Chadwick believes that in his third tractate on theology, Boethius

taught the Latin West, above all else, the method of axiomatization, that is, of analyzing an argument and making explicit the fundamental presuppositions and definitions on which its cogency rests. He taught his successors how to try to state truths in terms of first principles and then to trace how particular conclusions follow therefrom. The West learnt from him demonstrative method. (Chadwick 1981, 210)

Boethius began a trend that would eventually revolutionize Christian theology and transform it into a rationalistic and analytical discipline that made late medieval theology different from anything that followed in subsequent centuries, as we shall see in chapter 7. But if Boethius made dramatic changes in the way theology was done, he was also a Christian who never intruded anything into his secular works on logic and the quadrivium that might suggest he was a Christian. Most amazing of all is Boethius' behavior after he was condemned to death for treason by the emperor Theodoric in 524. As he awaited execution, Boethius wrote one of the great books in Western literature titled *On the Consolation of Philosophy*. In this final work, as he awaited death, Boethius gives no indication that he was a Christian. *On the Consolation of Philosophy* was a lofty, influential philosophical treatise, but one would much more readily infer that its author was a pagan than a Christian.

Cassiodorus, Isidore of Seville, and the Venerable Bede

Cassiodorus, Isidore of Seville, and the Venerable Bede were also important contributors to the modest storehouse of scientific knowledge that was available to those who lived during the low point of Western civilization. Author of numerous works, Cassiodorus founded a monastery called Vivarium. The primary objective of the monks at Vivarium "was to serve God by studying and copying the scriptures, the works of the Church Fathers, and the classics of antiq-

uity" (Thomas 1971, 109). The most famous of Cassiodorus' works is
the *Introduction to Divine and Human Readings,* divided into two books.
The first describes how to study scripture and emphasizes the utility
of the seven liberal arts in arriving at spiritual truth. The second book
is devoted to the seven liberal arts, with the greatest emphasis placed
on rhetoric and dialectic in the trivium and on arithmetic and music
in the quadrivium, with little attention paid to geometry and astronomy. Not surprisingly, Cassiodorus drew heavily on standard traditional handbook sources. His praise of the seven liberal arts and his
discussion of them encouraged many scholars in subsequent medieval
centuries to study those disciplines.

Isidore of Seville lived during the turn of the seventh century. He
was born into a prominent family in Roman Spain. In 599, Isidore suc-
ceeded his brother, Leander, as bishop of Seville. Isidore wrote extensively on scripture, theology, history, and science. He composed two
treatises relevant to science that were subsequently influential. The
earlier, titled *On the Nature of Things (De natura rerum),* is essentially
about cosmology and is divided into forty-eight chapters. Numerous
astronomical topics are discussed, among which are the parts of the
world, the planets, eclipses of sun and moon, the celestial waters, an
obviously biblical theme, and the course of the stars. Also included
are discussions of the earth and its parts, as well as various meteorological phenomena. *On the Nature of Things* was an encyclopedic work
about the physical world, covering themes that one could find discussed in the handbooks that had become integrated into the Roman
encyclopedic traditions best exemplified by Seneca and Pliny.

Much more famous than *On the Nature of Things* was Isidore's more
expansive encyclopedic treatise, the *Etymologies.* As one author puts
it, the *Etymologies*

briefly defines or discusses terms drawn from all aspects of human knowledge and is based ultimately on late Latin compendia and gloss collections.
The books of greatest scientific interest deal with mathematics, astronomy,
medicine, human anatomy, zoology, geography, meteorology, geology, min-
eralogy, botany and agriculture. Isidore's work is entirely derivative—he
wrote nothing original, performed no experiments, made no new observations or reinterpretations, and discovered nothing—but his influence in the
Middle Ages and Renaissance was great. (Sharpe 1973, 27)

Indeed, Isidore also treated the seven liberal arts, theology, and the
church, as well as numerous other topics. As its name implies, the *Etymologies* is a dictionary, but not presented in alphabetical order. Rather,

Isidore defines words within the context of a topic or theme and gives their meaning. Unfortunately, many of Isidore's word derivations are utterly absurd. In Book III, Isidore treats the four sciences of the quadrivium—arithmetic, geometry, music, and astronomy, in that order. All told, he devotes seventy-one very brief chapters to the four sciences, dealing with astronomy at greatest length (chapters 24–71) and drawing much of it from his earlier *On the Nature of Things*. Isidore titled chapter 29 "On the Universe and its name." In this one-paragraph chapter, Isidore says:

1. *Mundus* (the universe) is that which is made up of the heavens and earth and the sea and all the heavenly bodies. And it is called *mundus* for the reason that it is always in motion (*motus*). For no repose is granted to its elements. (Grant 1974, 12)

The derivation of *mundus* from *motus* is absurd.

In most chapters, Isidore gives tidbits of information about different aspects of the cosmos, primarily planets, especially the sun and moon, stars, and comets. Lynn Thorndike estimates that "Isidore contains less superstitious matter even proportionally to his meager content than Pliny does in connection with the virtues of animals, plants, and stones" (Thorndike 1923, 1:626). Although Isidore was an unoriginal compiler of facts and misinformation, his work was a source of "information" at a time when all forms of knowledge were in short supply. Isidore's *Etymologies* is a glaring reminder that learning on the continent of Europe had reached a deplorably low level. But it is also noteworthy that the secular knowledge, which Isidore describes in his two influential works, is presented in a way that fails to differentiate Isidore from any secular, pagan author in the Greco–Roman tradition. Where the theme is religious or scriptural, he will, of course, speak about these matters and use his knowledge of theology and religion to do so. But in topics and subjects that are about the physical cosmos and its many aspects, he simply uses relevant material that was available and does not attempt to use such knowledge as a stepping-stone for the introduction of religious or theological ideas.

The Venerable Bede was born in England and at the age of seven was given over to the abbot of Wearmouth monastery to be raised in the double monastery of Wearmouth and Jarrow, spending almost all of his life at Jarrow. "From that time," Bede wrote in 731 (in his *Ecclesiastical History*), four years before his death, "spending all the days of my life in residence at that monastery, I devoted myself wholly to

Scriptural meditation. And while observing the regular discipline and the daily round of singing in the church, I have always taken delight in learning, or teaching, or writing" (Jones 1970, 564). Fortunately for Bede, the library at the monastery was excellent for its time and thus enabled him to write treatises on science, history, and theology. In his numerous writings, Bede sought to emphasize the Christian religion. His most famous treatise is *The Ecclesiastical History of the English Nation*.

Bede also preserved, and added to, the science and natural philosophy of his day. As other Christian authors had done earlier, Bede wrote a work to which he gave the commonly used title *On the Nature of Things* (*De natura rerum*). This was compiled largely from the works of Pliny and Isidore of Seville and was intended as an introduction to the basic concepts of astronomy. "Bede's *On the Nature of Things*," in William Stahl's description, "is a concise textbook of fifty-one chapters, superimposing a Christian cosmography upon a framework of secular cosmographical and astronomical traditions" (Stahl 1962, 227–228). Although he used Pliny and Isidore, Bede usually treated their materials more intelligently than they did, but he occasionally fell victim to their errors, as when he agreed with Pliny that the moon is larger than the earth (Stahl 1962, 228). Bede followed Isidore and accepted a spherical world and a spherical earth. In addition to astronomical information, Bede also discussed meteorological topics, such as the seas, thunder, and comets.

Like so many other early Christian authors, Bede was interested in science as an aid to Christian understanding. Toward this end, he focused on calendar problems, especially the difficulty of determining the correct date for Easter. To cope with these problems, Bede wrote a treatise in 703 titled *On the Divisions of Time* (*De temporibus*). Because this work was so brief, it was difficult to follow, prompting Bede to revise it in 725 by expanding it to approximately ten times its former length. This became Bede's scientific masterpiece. William Stahl has identified three noteworthy features in Bede's great work:

Chapter 1 presents the main and almost unique account of early finger reckoning, a practice which was popular in the ancient world; and chapter 65 provides Easter tables for the years 532 to 1063. Chapter 29 contains the best extant account of tidal motions that had yet been written. Bede notes the differences between spring and neap tides and the effect of winds in retarding or advancing the flow of a tide. He writes as if from personal observation about the differences in times of tides at different points along the same shore. This phenomenon must have been common knowledge among fisherman and mariners, but, in any case, Bede's statement is the earliest record of the important principle of the "establishment of a port." (Stahl 1962, 232)

Bede's "establishment of the port" (or "establishment of a port") "has been described (e.g. by Duhem) as the only original formulation of nature to be made in the West for some eight centuries" (Jones 1970, 1:565).

WESTERN EUROPE AT ITS NADIR

There were other lesser authors who contributed to the overall stock of early science and natural philosophy. But those whose contributions I have briefly summarized were the most significant and represent the level of science that was attained during the fifth to eighth centuries. That it reached even this relatively low level of scholarship is rather remarkable when one considers the great societal changes that were underway as a result of the barbarian invasions that had begun in earnest in the fourth century and continued until the tenth century. From the fourth to the seventh centuries, the greatest danger came from the Germanic tribes who placed northern Europe in jeopardy. And when it seemed that Europe might settle down and consolidate in the eighth century under the rule of Charlemagne, his death in 814 ended the centralization tendencies that had been underway, and accelerated the emergence of feudal lords that splintered Europe into small principalities. As if that were not enough, the scourge of the Norsemen, or Vikings, began in the late eighth century and continued on into the tenth century. Over the course of the eighth to tenth centuries, the Vikings invaded at many places in Europe, including France (they besieged Paris in 885–886), the Low Countries, Britain, Ireland, and Spain. In France they came to be called Normans, and under that name invaded Sicily and southern Italy in the eleventh century; as warriors known as Rus, the Norsemen invaded Russia in the East and, at the least, left their name as a legacy.

The continuous series of barbarian invasions left Roman civilization in the West in a state of serious decline. The tradition of a centralized Roman government had virtually disappeared, replaced by fractionalized feudal kingdoms. Fortunately, the Germanic and Viking invaders settled down in Europe and merged with the native populations. Indeed, the Vikings were so thoroughly assimilated into the native populations of the lands they invaded that they left virtually no trace of their culture and language. As one author has described it, "after the thieving and the killing and the land-taking, they [the Norsemen] farmed and gradually became Englishmen, Irishmen, Scotsmen, Frenchmen, and Slavs" (Logan 1989, 436).

THE NEW EUROPE IN THE TWELFTH CENTURY

By the eleventh and twelfth centuries, the great intermingling of peoples and customs that had occurred in Western Europe from the fifth to the tenth centuries resulted in a new and vibrant Europe. Whatever may be the explanation for this extraordinary phenomenon, a new society emerged that differed dramatically from the Roman Empire, although it inherited much from that empire. From the standpoint of the relations between science and religion, one of the most noteworthy features of the new Europe was a remarkable emphasis on reason and rationality. Why this occurred is perhaps ultimately inexplicable, but the desire for reasoned explanations and analyses is a fact of the European society that emerged after the barbarian invasions. This is perhaps partially explicable by the fact that Europe benefited from a dramatic growth of its economy. A number of factors produced this state of affairs. Agricultural output increased dramatically because of new technological advances, such as the heavy plough and the use of horses, rather than oxen, to pull them. This, in turn, was a development that occurred by virtue of the introduction of the nailed horseshoe and the horse collar. The horse collar, padded and rigid, did not choke the horse, unlike the yoke harness it replaced—which often caused death by suffocation—and even enabled the horse to pull a load four or five times greater than with the yoke harness. Along with the replacement of the two-field system of crop rotation by the three-field system, which enabled farmers to increase their productivity by one-third, these advances generated a great increase in agricultural output.

With greatly increased food supplies, the population grew dramatically. Europeans founded hundreds of new towns and cities and began to populate areas of Europe, especially in the east, that had previously been only lightly populated. By the end of the twelfth century, the level of commerce and manufacturing was greater than it had been at the height of the Roman Empire (see Grant 2001, 20–21). After centuries of decline, cities became a powerful force in the cultural, religious, economic, and political life of Western Europe. Between the tenth and thirteenth centuries, Europe developed a money economy driven by the commercial activities of the new cities.

As the economy grew and the cities developed, relations between church and state were also significantly altered. From the outset of Christianity, it was assumed (as we shall see in chapter 8) that church and state were separate from one another, and each was supreme in its own domain. But throughout the early Middle Ages, each sought to dominate the other whenever an opportunity presented itself. From the sixth to the eleventh centuries, however, it was the state that imposed its will on the church. Secular rulers—kings and powerful nobles—often appointed the higher clergy—bishops and abbots. Indeed, the appointees were often relatives of the secular rulers. Moreover, secular rulers often assumed the authority to invest new church appointees with a double investiture, ecclesiastical and temporal, when they should only have conferred the temporal investiture and left the ecclesiastical investiture to a high church official.

All this changed in the first quarter of the eleventh century by virtue of the Investiture Struggle (A.D. 1075–1122), which began with Gregory VII, who was pope from 1073 to 1085. The Investiture Struggle was a conflict between church and secular authorities over control of church offices begun by Gregory when he proclaimed supremacy of the church over the state in a document he wrote in 1075, known as the The Pope's Dictate (Dictatus Papae). In that document, Gregory asserted, among numerous similar statements, that "of the pope alone all princes shall kiss the feet" (Thompson and Johnson 1937, 379), and that he had the authority to depose emperors.

What Gregory VII began culminated in 1122 at the Concordat of Worms under the pontificate of Calixtus II, a French pope. Under the terms of the Concordat, the Holy Roman Emperor agreed to take no part in spiritual investiture and to allow free ecclesiastical elections. Because the consequences of this agreement were of great significance, scholars have come to label the subsequent period of the papacy as the Papal Revolution (Berman 1983, 85–119). Freed from domination by secular rulers and officials, the papacy was now in control of its own clerical members. It quickly evolved into a powerful, centralized bureaucracy. Of the greatest importance in this momentous development was the fact that literacy was concentrated in the church. This was a powerful tool for exercising its power. Harold Berman declares that the Papal Revolution "may be viewed as a motive force in the creation of the first European universities, in the emergence of theology and jurisprudence and philosophy as systematic disciplines, in the creation of new literary and artistic styles, and in the development of a new consciousness" (Berman 1983, 100). We shall see how the church came to grips with secular learning and a growing reliance on reason and reasoned argument. Indeed, that history begins even before the Papal Revolution.

From the eighth to the tenth centuries, the kind of emphasis on reason that Boethius exhibited made an appearance in a few monasteries and in the courts of Charlemagne and Charles the Bald.

Reason and the Challenge to Authority: John Scotus Eriugena

In the ninth century, John Scotus Eriugena (c. A.D. 810–d. after 877), the Irish philosopher and theologian, who served King Charles the Bald of France, wrote a significant treatise, titled *On the Division of Nature*. In this work, Scotus Eriugena reveals himself as one of the earliest medieval authors to exhibit the new tendency to rely on reason, and even to exalt it, as we see in this declaration:

For authority proceeds from true reason, but reason certainly does not proceed from authority. For every authority which is not upheld by true reason is seen to be weak, whereas true reason is kept firm and immutable by her own powers and does not require to be confirmed by the assent of any authority. For it seems to me that true authority is nothing else but the truth that has been discovered by the power of reason and set down in writing by the Holy Fathers for the use of posterity. (Abelard 1979, 82–83, n. 136)

In this extraordinary claim, Scotus Eriugena subordinates authority to reason. This is noteworthy because it marks the earliest beginning of a powerful trend in later medieval thought, a trend that was directly relevant to the long-term relations between science and religion. The challenge to authority was ultimately a challenge to all authority: to the authority of Greco–Roman science and natural philosophy and the authority of the church fathers. This should not be taken to mean that authority was rejected during the Middle Ages—far from it. But appeals to reason become more frequent, and the nature of the medieval university curriculum, as we shall see, was such as to encourage this tendency. There was, however, a limit to the application of reason; reason could not be applied to revealed truths—for example, the doctrines of the Trinity and the Eucharist, which were regarded as beyond the understanding of reason.

John Scotus Eriugena's emphasis on reason was given institutional roots in eleventh-century Europe with the development of the cathedral schools that emerged in various European cities. In the late eighth century, Charlemagne decreed that cathedrals and monasteries were to establish and support schools to educate boys, with the hope that it would provide educated priests for the church. Viking invaders and other intruders into Charlemagne's Holy Roman Empire greatly hindered the implementation of this decree. In 1079, however, Pope Gregory VII issued a papal decree that ordered cathedrals and monasteries to establish schools to educate and train priests. Gregory's decree had a significant impact. Cathedrals, which were located in population centers, were thereafter expected to provide their clergy and others with free instruction in Latin and in the liberal arts. By the eleventh century, the cities of Paris, Orleans, Toledo, Rheims, Chartres, Cologne, and others had cathedral schools that had become intellectual centers for both clergy and non-clergy. By this time, there were large numbers of students in Europe who were eager for an education and were willing to travel long distances and endure many hardships to obtain it. At cathedral schools, they found libraries and other students who shared their desire for an education.

The Emergence of Logic in Medieval Education

The resurrection of Boethius' old logic was a major impetus behind the application of reason to all kinds of problems in the late tenth century. Although the old logic had been around since the sixth century, few used it; and it was not really taught in schools. But in the late tenth century, Gerbert of Aurillac (c. A.D. 946–1003), who became Pope Sylvester II (A.D. 999–1003), gained great fame as a teacher of the seven liberal arts at the cathedral school of Rheims. He had a high regard for logic, or dialectic, as it was often called, and may have been the first to focus attention on the numerous treatises that comprised Boethius' old logic. In the eleventh century, Gerbert's students disseminated his love of learning and his teaching methods throughout northern Europe. As a consequence, logic became a basic subject of study in the cathedral schools of Europe. And, in the twelfth and thirteenth centuries, would become ever more deeply entrenched in the curricula of the cathedral schools and then the universities of Europe. Why did this happen? Why should the forbidding subject of logic play such a central role in medieval education? R. W. Southern has asked, "why it was that, from the time of Gerbert, this study assumed an importance which it had never previously attained in the Latin world. The works of Boethius are immensely difficult to understand and repellent to read. Why should the subject have taken such a hold on the imaginations of scholars, so that they pursued it with unflagging zeal through all the obscurities of translation, heedless of the advice of many cautious men of learning?" (Southern 1953, 179).

Perhaps logic became a major subject of study in the eleventh century because it "opened a window on to an orderly and systematic view of the world and of man's mind" (Southern 1953, 179) at a time when intellectual life was at a low ebb. Subjects like theology and law were difficult to study because, over the centuries, they had become riddled with inconsistencies and contradictions. The old logic of Boethius served the early Middle Ages as a model of rigor and organization for the floundering disciplines of law and theology, and perhaps medicine as well. As difficult a subject as it was, logic became a fixture in medieval education. Hugh of St. Victor (d. A.D. 1141) spoke for most in the Middle Ages when, in his widely read *Didascalicon*, he declared "logic came last in time, but is first in order. It is logic which ought to be read first by those beginning the study of philosophy, for it teaches the nature of words and concepts, without both of which no treatise of philosophy can be explained rationally" (Hugh of St. Victor 1961, 59).

John of Salisbury (c. A.D. 1115–1180) reinforced Hugh of St. Victor's praise of logic. John had studied under some of the great masters of his day, including Peter Abelard (see below). He was not only well acquainted with the seven liberal arts, but was also learned in theology. John was an ordained priest who witnessed the murder of Thomas Becket on December 29, 1170. In 1176, he was made bishop of Chartres, where he died on October 25, 1180. John of Salisbury was the author of numerous works, but his most relevant treatise about education in general, and the seven liberal arts in particular, was his *Metalogicon*, in which he gave an assessment of logic. John left no doubt that he was a strong believer in the use of reason and that logic was its powerful instrument. He observed that Saint Augustine had praised logic so "that only the foolhardy and presumptuous would dare to rail against it." Indeed, "since logic has such tremendous power," John continues, "anyone who charges that it is foolish to study this [art], thereby shows himself to be a fool of fools" (John of Salisbury 1955, 241). Logic came into being, says John, because

there was evident need of a science to discriminate between what is true and is false, and to show which reasoning really adheres to the path of valid argumentative proof, and which [merely] has the [external] appearance of truth, or, in other words, which reasoning warrants assent, and which should be held in suspicion. Otherwise, it would be impossible to ascertain the truth by reasoning. (John of Salisbury 1955, 76)

John argues that logic is to philosophy as the soul is to the body. Without the organizing principle of logic, all philosophy "is lifeless and helpless." And yet, despite his enthusiasm for logic, John also emphasized its limitations. It was not to be studied as an end in itself, but only to better understand other subjects. Most importantly, logic and reason should not be used to elucidate divine mysteries, such as the Trinity. This would be to reach beyond our level of comprehension. In this sense, John of Salisbury reflects the more traditional attitude of scholars and theologians.

Where John of Salisbury chose to reject the application of logic and reason to articles of the faith, others followed a different path, one that eventually made the application of logic and reason to the divine mysteries a common occurrence. The apparent power of logic and reasoned argumentation encouraged some early medieval authors to apply these powerful tools to theology, especially to the revealed truths of the faith, which were the focal points of medieval Christian theology. The application of reason and logic to revelation, and to theology in general, is a significant part of the medieval struggle between faith and reason, or, in its broader context, the relations between science and religion. There were indeed other important facets of the relations between science and religion in the late Middle Ages, but the interaction of reason and revelation appeared even before the introduction of Aristotle's natural philosophy in the mid-twelfth century, and the subsequent transformation of medieval natural philosophy. We already saw the attitudes of various church fathers to the application of secular science for the better understanding of scripture. In the eleventh century, a few theologians deliberately sought to apply reason to revealed truths that were usually regarded as beyond the powers of human reason. Although the application of reason to revelation had its definite limits, the use of reason in theology was regarded as an important instrument for the proper understanding of revelation. In this sense, reason came to be opposed to authority, in the manner advocated by John Scotus Eriugena. Some church officials, however, came to regard the rush to reason as a dangerous trend and vigorously opposed it.

The Role of Reason in Theology: Berengar of Tours, Anselm of Canterbury, and Peter Abelard

Berengar of Tours and Anselm of Canterbury show that already in the eleventh century reason was taking root as an instrument of theology. Berengar of Tours (c. A.D. 1000–1088) believed that evidence was more important than authority in theology, and he was convinced that reason, which he regarded as a gift of God, should be applied to faith. In his explication of the Eucharist, or doctrine of transubstantiation, Berengar clashed with traditional theology. Using the old logic, he insisted that reason could not support the view that accidents can exist apart from a substance, as required by church doctrine, which holds that the accidents of the bread no longer subsist in it after the consecration; that is, because the bread no longer exists, its accidents can no longer inhere in it, but must exist independently. Berengar found this illogical and assumed that the form of the bread continued to exist with its accidents still inhering in it. He then assumed that a second form, which is the body of Christ, is added to the form of the bread.

Berengar clashed with Lanfranc of Bec (A.D. 1010–1089), an Italian who died as archbishop of Canterbury. Lanfranc was not an opponent of logic but, like John of Salisbury, opposed its application to the mysteries of the faith. Lanfranc criticized Berengar for relying on reason and ignoring sacred authorities in his analyses, to which Berengar replied that "it is incomparably superior to act by reason in the apprehension of truth; because this is evident, no one will deny it except a person blinded by madness" (Holopainen 1996, 109). In the same rebuttal of Lanfranc, Berengar declared that "to have recourse to dialectic is to have recourse to reason; and he who refuses this recourse, since it is in reason that he is made in the image of God, abandons his glory, and cannot be renewed from day to day in the image of God" (Holopainen 1996, 116). Berengar accepted traditional writings about

Christian doctrine upheld by the church, but he strongly believed that the use of reason was essential for understanding them properly.

Thus, both Berengar and Lanfranc regarded logic, or dialectic, as important, but the former was prepared to apply it to articles of the faith, while the latter was not. Those who had learned and used logic were divisible into two groups: those who were prepared to use logic to study articles of the faith, and those who were not because they regarded it as improper "to submit the mysteries of faith to the rules of logical reasoning" (Gilson 1955, 615, n. 41). In general, it seems that "to each theologian, the proper use of dialectics was the one which he himself was making of it" (Gilson 1955, 615, n. 41).

More than anyone else, Anselm of Canterbury (A.D. 1033–1109) advanced the cause of those who wished to apply reason to theological problems. Following a sojourn as prior and then abbot of the Abbey of Bec, Anselm became archbishop of Canterbury in 1093. Along with many like-minded theologians, Anselm firmly believed, as had Saint Augustine, that in order to understand the faith, one had first to be a believer. Reason presupposed faith to gain understanding. Anselm thought it essential to explain in rational terms what he believed on faith. He wrote two major theological treatises, the *Monologium* and the *Proslogium*. The three proofs of God's existence in the former treatise and the famous ontological proof in the latter are based solely on reason and do not involve appeals to scripture or revelation. The *Monologium* is a remarkable treatise because, as Anselm informs us, it was requested by Anselm's fellow monks with whom he had discussed the being of God and other topics. The form of their request is, however, quite unusual. In Anselm's words, they asked "that nothing in Scripture should be urged on the authority of Scripture itself, but that whatever the conclusion of independent investigation should declare to be true, should, in an unadorned style, with common proofs and with a simple argument, be briefly enforced by the cogency of reason, and plainly expounded in the light of truth" (Anselm 1944, 35). Thus, the monks requested that Anselm avoid appeals to scripture and use only reason to demonstrate truths of faith, especially the existence of God. In both treatises, Anselm used only reason, arguing as if he had never heard of Christianity. The most famous example of Anselm's use of reason is his controversial ontological proof. Here is a modern philosopher's summary of it:

The definition of God in Whom all Christians believe, contains the statement that God is a Being than which no greater can exist. Even the Fool in the Psalm (Ps.14:1), who said that there was no God, understood what was meant by God when he heard the word, and the object thus defined existed in his mind, even if he did not understand that it exists also in reality. But if this being has solely an intra-mental existence, then another can be thought of as having real existence also, that is, it is greater (by existence) than the one than which no greater can exist. But this is a contradiction in terms. Therefore the Being than which no greater can be conceived exists both in the mind and in reality. (Knowles 1962, 103)

Down to the present day, philosophers have argued about the validity and meaning of Anselm's ontological proof for the existence of God, thus testifying to its importance.

Anselm's treatment of theology was rigorous enough that modern historians properly regard him as having laid the foundation for the subsequent transformation of theology into a science in the twelfth and thirteenth centuries. A historian of theology has compared the attitudes of Berengar of Tours and Anselm of Canterbury towards reason. Both were champions of reason, but "for Berengar, the primary task of reason in theology is to function as a means of interpreting the authoritative writings of the Church. For Anselm, the primary task of reason in theology is to construct rational demonstrations for articles of faith" (Holopainen 1996, 132). Thus did Anselm place even greater emphasis on reason than did Berengar.

The most significant logician and the greatest rationalist of the twelfth century was Peter Abelard (see Figure 5.1), perhaps the most important intellectual of the twelfth century, and forever famous for his love affair with Heloise. Unlike Berengar of Tours, who used logic to challenge the doctrine of transubstantiation, Peter Abelard, like Anselm and most other advocates of the application of reason to theology, did not use reason to challenge church doctrine, as he indicates in this statement: "I will never be a philosopher, if this is to speak against St. Paul; I would not be an Aristotle, if this were to separate me from Christ" (Knowles 1962, 123). Like Anselm, Abelard was also convinced that one should accept divine authority as supreme, but that in order to understand the faith it was necessary to use logic and reason. He argued that philosophy was essential for the proper defense of the faith against heretics and unbelievers. "Those who attack our faith," Abelard declared in one of his theological treatises, "assail

Figure 5.1. Peter Abelard.

us above all with philosophical reasonings. It is those reasonings which we have principally enquired into and I believe that no one can fully understand them without applying himself to philosophical and especially to dialectical studies" (Luscombe 1988, 294, n. 57).

With this attitude, Abelard applied logic and reason to the articles of faith. In his autobiography, *The Story of My Misfortunes* (*Historia calamitatum*), Abelard explains that he wrote a theological treatise on the divine unity and trinity for his students, "who were asking for human and philosophical reasons and clamouring more for what could be understood than for what could be said." His students were apparently agreed that "the utterance of words was superfluous unless it were followed by understanding, and that it was ridiculous for anyone to preach to others what neither he nor those taught by him could accept into their understanding" (Luscombe 1988, 293).

Abelard was driven to understand whatever he was asked to accept, and he wanted his students and others to do the same. To illustrate his great desire to study all questions in theology from all relevant aspects, Abelard wrote his most famous treatise, *Yes and No* (*Sic et Non*). Because he wanted his students to approach all questions in a critical spirit, Abelard presented a series of theological questions in which alternative answers were presented for each question. In order to elicit conflicting responses to the questions, Abelard framed each question in such a way that both yes and no answers were expected. He usually ended each question with "and the contrary" (*et contra*) or "and not" (*et non*). Among the questions he considered, in their original numbering, are

1. That faith is to be supported by human reason, *et contra*.
32. That to God all things are possible, *et non*.
141. That works of mercy do not profit those without faith, *et contra*.
145. That we sin at times unwillingly, *et contra*.
154. That a lie is permissible, *et contra*.
157. That it is lawful to kill a man, *et non*. (Haskins 1957a, 354–355)

In his response to each question, Abelard presented the most important affirmative and negative arguments, drawing most of them from the church fathers. In presenting arguments on both sides of each question, Abelard drew attention to the many differences of opinion among the church fathers. One could see at a glance that they were often in serious disagreement with one another. Many thought Abelard had subverted the church by focusing on such discrepancies. After all, if the church fathers could not agree on many of these significant doctrinal questions, what were the faithful to think? Abelard explains his intent at the conclusion of his prologue to *Yes and No*:

The discrepancies which these texts seem to contain raise certain questions which should present a challenge to my young readers to summon up all their zeal to establish the truth and in doing so to gain increased perspicacity. For the prime source of wisdom has been defined as continuous and penetrating enquiry. The most brilliant of all philosophers, Aristotle, encouraged his students to undertake this task with every ounce of their curiosity. In the section on the category of relation he says: 'It is foolish to make confident statements about these matters if one does not devote a lot of time to them. It is useful practice to question every detail.' By raising questions we begin to enquire, and by enquiring we attain the truth, and, as the Truth has in fact said: 'Seek, and ye shall find; knock, and it shall be opened unto you.' . . . When, therefore, I adduce passages from the scriptures it should spur and incite my readers to enquire into the truth and the greater the authority of these passages, the more earnest this enquiry should be. (Piltz 1981, 82)

In this powerful and significant paragraph, Abelard emphasizes rational investigation with his declaration that "by raising questions we begin to enquire, and by enquiring we attain the truth." Abelard was acutely aware of the pitfalls in the pursuit of truth by rational means. In the same prologue,

Abelard discusses how to compare what different writers say and how to account for the meaning of a word, which may take on different meanings in different contexts. It is essential to interpret troublesome texts and passages and to determine, to the extent possible, whether they are the result of miscopying or poor translation, or due to the reader's ignorance. If, after careful analysis, two authorities still disagree, it is necessary to weigh one against the other in the larger scheme of things. Nothing is exempted from this rigorous inspection except the Bible and those pronouncements that the Church has accepted as true. All other authorities and texts are open to criticism and analysis in an effort to arrive at the truth. (Grant 2001, 61)

Abelard was perhaps the most spectacular author in the twelfth century to rely on reason and logic in his analysis of theology and its claims and arguments. The twelfth century proved a significant period for the advance of theology. But it did not come easily.

The Reaction to the Use of Reason in Theology

Not all theologians were pleased to witness the foundations of the new scholastic theology being laid in the late eleventh and twelfth centuries. In the eleventh century, Saint Peter Damian (A.D. 1007–1072) reacted harshly to the new trend. He was hostile to the seven liberal arts and wholly disapproved of the application of logic to theology and the faith. As a sign of his contempt for logic, Peter allowed that God could undo a historical event by simply willing that it had not happened. Although such a divine action would produce a logical contradiction and make logical contradictions a part of logic, a step that all logicians would repudiate as absurd, Peter found such a move quite compatible with God's infinite power.

The defenders of traditional theology, who were largely associated with monasteries, emphasized contemplation rather than analysis and were therefore suspicious of the new theology, which seemed to encourage curiosity and much poking around into all aspects of every theological question. Typical of the traditional attitude was Rupert of Deutz (A.D. 1070–c. 1129), a monastic theologian who was famous in his day. Rupert denounced the new theologians in the following terms: "Shamefully, they dared to examine the secrets of God in the Scriptures in a presumptuous way, motivated by curiosity and not by love. As a result they became heretics. God has decreed that the proud are not to be admitted to the sight of divinity and truth" (Gaybaa 1988, 43).

Other monastic scholars hostile to the new theology can be cited, but the most significant by far was Saint Bernard of Clairvaux (A.D. 1090–1153). Bernard was not only hostile to the new theology, especially to the application of logic to theology, but he also found the new emphasis on the liberal arts distasteful. Bernard found the teachings of Peter Abelard particularly offensive. As a powerful theologian, Bernard used his influence to attack Abelard whenever he could. He sent many letters to church authorities, including Pope Innocent II, denouncing the ideas of Abelard. It was Abelard's excessive reliance on reason that worried Bernard, especially his application of it to theology. In a letter to a cardinal, Bernard said of Peter:

He has defiled the Church; he has infected with his own blight the minds of simple people. He tries to explore with his reason what the devout mind grasps at once with a vigorous faith. Faith believes, it does not dispute. But this man, apparently holding God suspect, will not believe anything until he has first examined it with his reason. (Bernard of Clairvaux 1953, 328)

Bernard's continuous attacks against Abelard resulted in a charge of heresy against Abelard, who was condemned at the Council of Sens in 1141. Abelard was compelled to forego lecturing, and he retired to the monastery of Cluny, where he remained until his death.

The *Sentences* of Peter Lombard

Although Peter Abelard and others were silenced by conservative theologians, who still had influence and power to affect the course of theology, the days of the traditional theologians were numbered. The application of reason and logic to theology was too important to be stifled for long. Shortly after the death of Bernard of Clairvaux, theology received a major, even monumental, impetus from the composition of a theological text by Peter Lombard (c. A.D. 1095–1160; see Figure 5.2). Between 1155 and 1158, Peter Lombard wrote a treatise titled *Four Books of Sentences*, which became the basic textbook in all schools of theology in the Latin West until the seventeenth century. The four books into which Peter divided his treatise were devoted to God, the Creation, the Incarnation, and the sacraments, respectively. Theological students were required to lecture and comment upon these four books. The term *sentences* in the title stands for the Latin word *sententiae*, or opinions. The opinions were drawn from the writings of the church fathers. Peter's object was to organize these opinions as logically as possible, thus forming a new "systematic theology." The church father upon whom Peter relied most heavily was Saint Augustine. Peter was not the first to write a theological textbook, but his collection of opinions displaced all others and became the standard work for theologians to study and master. Between 1150 and 1500, only the Bible was read and discussed more than the *Sentences*.

Because the treatise was lectured and commented upon by all theology students, there is a very large body of extant commentaries on the *Sentences*, perhaps as many as 1,000, many of great length. Examination of this body of literature indicates unequivocally that medieval theology was a systematic, rationalistic discipline. It also reveals that a great deal of natural philosophy was introduced into these com-

De A. Theuet, Liure III.   142

PIERRE LOMBARD MAISTRE
des Sentences.   Chap. 69.

Figure 5.2.   Peter Lombard. (New Catholic Encyclopedia, 2003, 11:191.)

mentaries. This occurred largely because the second book was devoted to the Creation, making it almost inevitable that commentators would use Aristotle's natural philosophy—especially his physics and cosmology—to explain the Biblical creation. But logic was also extensively applied in all four books. How commentaries on the *Sentences* illustrate the relationship between science and religion, or natural philosophy and theology, will be discussed in chapter 7.

The victory of the new theology was guaranteed with the introduction into Europe of Latin translations of Greco–Arabic science and
natural philosophy. Within this body of literature, Aristotle's logic and
natural philosophy played the crucial role that transformed both natural philosophy and theology.

THE BEGINNINGS OF THE NEW NATURAL
PHILOSOPHY

As theology changed in the first half of the twelfth century, so did
natural philosophy. Dramatic changes began even prior to the introduction of Aristotle's works, which impacted medieval thought in the
second half of the twelfth century and had their full impact in the thirteenth century. The changes in natural philosophy prior to the transmission of Greco–Arabic science into the Latin West were parallel to
those that occurred in theology, and which were just described. That
is, natural philosophers began to question authority; they sought to
interpret nature in rational terms and accepted traditional authorities
only if they passed the test of reason. The two most significant twelfth-
century natural philosophers were Adelard of Bath (c. A.D. 1080–1142)
and William of Conches (d. after A.D. 1154).

Adelard of Bath traveled in the Middle East and learned enough Arabic to translate Euclid's *Elements* and al-Khwarizmi's astronomical tables from Arabic to Latin. The most important of Adelard's works in
natural philosophy is the *Natural Questions* (*Quaestiones naturales*), written around 1116. The treatise is in the form of a dialogue between Adelard and his unnamed nephew. The nephew, who often makes foolish
statements, poses question after question to Adelard, who patiently answers them, as a good uncle should. The questions range over plants,
animals, the elements, man, physics, meteorology, and astrology. Adelard compares authority to being led around by a halter: "For what
should we call authority but a halter? Indeed, just as brute animals are
led about by a halter wherever you please, and are not told where or
why, but see the rope by which they are held and follow it alone, thus
the authority of writers leads many of you, caught and bound by
animal-like credulity, into danger" (Dales 1973, 41). He denounces
gullible, contemporary readers, "who require no rational explanation
and put their trust only in the ancient name of a title" (Dales 1973, 41).

After his nephew sees a flaw in one of Adelard's explanations about the nourishment and sustenance of plants and concludes that only an appeal to God can explain the phenomenon, Adelard replies to his nephew by invoking reason:

I take nothing away from God, for whatever exists is from Him and because of Him. But the natural order does not exist confusedly and without rational arrangement, and human reason should be listened to concerning those things it treats of. But when it completely fails, then the matter should be referred to God. Therefore, since we have not yet completely lost the use of our minds, let us return to reason. (Dales 1973, 40)

For Adelard, reason was a gift of God and should be used, or it would have been given in vain. God created our world and made it operate and function in a rational manner. The natural philosopher's duty is to seek rational explanations for natural phenomena. To those who were serious about investigating nature Adelard advises, "first we ought to seek the reason for anything, and then if we find an authority it may be added. Authority alone cannot make a philosopher believe anything, nor should it be adduced for this purpose" (Dales 1973, 41–42). Few in the Middle Ages surpassed Adelard's zeal for the use of reason in natural philosophy. He avoided appeals to divine power and sought a rational explanation for all phenomena.

William of Conches was a teacher in France but gave it up after becoming disillusioned with the pandering tactics of his colleagues. William's views on natural philosophy appear in his most important treatise, *Dragmaticon*, which was a revision of an earlier treatise titled *Philosophy of the World* (*Philosophia mundi*). For the first version of his treatise, William was severely criticized by William of St. Thierry, a friend and colleague of Bernard of Clairvaux. William of St. Thierry was incensed about the way William of Conches had discussed the Trinity and accused him of heresy. In the later *Dragmaticon*, William of Conches yielded to the criticism and made appropriate changes. William ranges over many topics in natural philosophy, but it is his attitude toward nature, and the role of the Bible in interpreting nature, that makes the *Dragmaticon* an important treatise.

Like Adelard of Bath, William of Conches did not recognize the church fathers as authorities in natural philosophy. "In those matters that pertain to the Catholic faith or moral instruction," William explains, "it is not allowed to contradict Bede or any other of the holy fathers. If, however, they err in those matters that pertain to physics, it is permitted to state an opposite view. For although greater than we, they were only human" (William of Conches 1997, 38–39). In agreement with Adelard, William firmly believed that God had created a rational universe that operated in a lawful manner. It was the task of the natural philosopher to discover the causes by which the universe operated. In another treatise, *Gloss on Boethius*, William denies the utility of scripture in natural philosophy. He argues that some priests believe only what they find in the Bible and ridicule topics and themes that are absent from sacred texts. "They don't realise," William explains, "that the authors of truth are silent on matters of natural philosophy, not because these matters are against the faith, but because they have little to do with the strengthening of such faith, which is what those authors are concerned with. But modern priests do not want us to inquire into anything that isn't in the Scriptures, only to believe simply, like peasants" (Stiefel 1985, 86).

With William of Conches and Adelard of Bath, natural philosophers began to subject their interpretations to the test of reason and logical analysis. Traditional authorities had to pass the test of reason before their explanations were accepted. Because they rarely invoked scriptural authority, the new natural philosophers did not include appeals to the Bible. There was little of religious content in their treatises about the workings of nature.

The historical developments described in this chapter may be viewed as a two-track evolution that challenged authority in two different, but sometimes related, domains. One track is represented by theology, to which the application of reason is limited only by revealed truths, which were beyond the scope of reason; the other track was that of natural philosophy, which was limited by certain fundamental truths associated with the biblical creation account, but not by revealed truths, with which it was not concerned.

The two tracks were destined for dramatic changes produced by momentous events in the twelfth century that changed the character and depth of natural philosophy in ways that came to seriously affect both theology and natural philosophy and their interrelations. The great weakness of twelfth-century natural philosophy—the kind of philosophy produced by Adelard of Bath and William of Conches— was its shallow and meager base of knowledge. But even as Adelard and William were writing their treatises, a massive change was underway that would transform Europe's intellectual culture. European scholars began to translate Greek and Arabic treatises into Latin, treatises that were previously unknown, or known only by the names of their eminent authors. At the core of this new literature lay the body of Aristotle's massive corpus of logic, science, and natural philosophy, a body of literature that would form the basis of natural philosophy for the next 500 years. To understand how this happened, we must describe the events that made it possible.

Chapter 6

━ ⊨◊⊨ ━

The Medieval Universities and the Impact of Aristotle's Natural Philosophy

By the end of the eleventh century, Christian Europeans were well aware that the Arabs in Spain and Sicily had attained a far higher level of knowledge and expertise in natural philosophy and the sciences than they had. Although they were undoubtedly aware of this long before the end of the eleventh century, their awareness was heightened immeasurably after they captured Toledo and Sicily from the Muslims in 1085 and 1091, respectively. Because both Toledo and Sicily were important centers of learning, Christians could directly witness the great disparity between their level of knowledge about the physical world and that of the Muslims whom they had just conquered. They also knew that Greeks in the Byzantine Empire and in various parts of Europe, especially Venice, had philosophical and scientific treatises in the Greek language that were unknown in Western Europe. As word spread into all parts of Europe, scholars were drawn to both Greek and Arabic centers of learning to translate these texts into Latin, the language of learning in Western Europe. As a consequence, a massive translation of Greek and Arabic science into Latin began in the twelfth century and continued until the third quarter of the thirteenth century.

Translations from Greek into Latin were preferred simply because Greek and Latin are cognate languages, and a more precise rendering could be made. Because Latin and Arabic are unrelated languages, misinterpretations were more likely to occur. But the essential objective for all translators was the same: to transfer faithfully the meaning of the text from one language to another. How successful were medieval translators? David Lindberg has rightly declared that

it is not possible to say much more than that translations varied from good to bad. Some translators, skilled in the subject matter as well as the relevant languages, were able to capture exactly the sense of the original; others, having an imperfect command of the languages, or the subject matter, produced unintelligible results. Viewed as a whole, however, translations provided Western Christendom with an adequate knowledge of the Greek and Arabic intellectual achievement—and thus with the basic materials out of which its own system of philosophy and natural science would be constructed. (Lindberg 1978, 79)

Although many scientific works in optics, astronomy, mechanics, mathematics, and alchemy were translated in the first half of the twelfth century, these works did not play a significant role in the relations between science and religion during the late Middle Ages. It was Aristotle's works on natural philosophy, and the commentaries on those works by his great Muslim commentator, Averroes (Ibn Rushd), that were, and remained, the focal point in the relations between science and religion. The translations of Aristotle's works were therefore of fundamental importance for the history of science and religion in the Middle Ages.

THE TRANSLATIONS OF ARISTOTLE'S BOOKS ON NATURAL PHILOSOPHY

Beginning in the twelfth century, Aristotle's works were translated from both Greek and Arabic. Translations from the Greek were preferred because it was the language in which Aristotle had written his treatises. The two greatest translators from Greek to Latin were James of Venice (d. after A.D. 1147), the first major translator of Aristotle's works from Greek to Latin, and William of Moerbeke (c. A.D. 1215–c. 1286), who was the last. James of Venice translated Aristotle's *Physics*, *On the Soul*, the *Parva naturalia*, and at least part of the *Metaphysics*. He also translated the *Posterior Analytics* and other logical treatises (Dod 1982, 54–55), which were regarded as the genuine Aristotelian texts for more than three centuries (Minio-Paluello 1973, 66).

William of Moerbeke translated at least forty-eight treatises, including seven on mathematics and mechanics by Archimedes, translated for the first time into Latin (Grant 1974, 39–41; Minio-Paluello

1974, 436–438). His Aristotelian translations are truly impressive. He was the first to translate Aristotle's biological works from Greek into Latin. In translating the rest of Aristotle's natural philosophy, Moerbeke found it useful to revise, expand, and even complete some earlier translations, including revisions of at least three treatises previously translated by James of Venice. In addition, Moerbeke translated Greek commentaries on Aristotle's works from late antiquity. Thus, he translated John Philoponus' *Commentary on the Soul*, and Simplicius' *Commentary on the Heavens*. One of the earliest beneficiaries of Moerbeke's translations was Thomas Aquinas, who used the translations of Aristotle's *On the Soul* and *Physics*, Simplicius' *Commentary on On the Heavens*, Themistius' *Commentary on the Soul*, and others (Weisheipl 1974, 152). With Moerbeke's monumental contributions, all of Aristotle's natural philosophy was available by the last quarter of the thirteenth century in translations from Greek and Arabic.

Although many scientific works were translated from Arabic to Latin in the first half of the twelfth century by such translators as Plato of Tivoli, Adelard of Bath, Robert of Chester, Hermann of Carinthia, Dominicus Gundissalinus, Peter Alfonso, John of Seville, and others, the earliest translations of Aristotle's works on natural philosophy appear to have occurred in Spain in the latter half of the twelfth century. By far the most prominent translator of Aristotle's natural philosophy was Gerard of Cremona (c. A.D. 1114–1187), the most prolific translator from Arabic to Latin of works on science, medicine, and natural philosophy. From a brief bio-bibliography that his students appended to Gerard's translation of one of Galen's medical treatises (*Tegni* or *Ars parva*), we learn that Gerard, unable to locate a copy of Ptolemy's *Almagest* among the Latins, went to Toledo in search of that great astronomical treatise. Once there, Gerard not only found the *Almagest*, which he eventually translated, but, his students inform us, "seeing the abundance of books in Arabic on every subject, and regretting the poverty of the Latins in these things, he learned the Arabic language, in order to be able to translate" (Grant 1974, 35). And translate he did. In approximately thirty years, he converted seventy-one works from Arabic to Latin. What is remarkable about Gerard's monumental effort is the range of his achievements, which included works in mathematics, astronomy, medicine, and alchemy; under the rubric of philosophy, he also translated most of Aristotle's natural philosophy. Included among his translations are Aristotle's *Physics*, *On the Heav-*

ens (*De caelo*), *On Generation and Corruption, Meteorology,* Books I–III (the fourth had already been translated). Gerard also translated the *Posterior Analytics,* Aristotle's important work on the theory of science.

Another extremely important translator from Arabic to Latin was Michael Scot, who not only translated important treatises in Aristotle's natural philosophy, but also translated some of the major commentaries on Aristotle's books by the great Islamic commentator Averroes, or Ibn Rushd (A.D. 1126–1198), who lived and worked in Spain. Of all Islamic authors whose works were translated into Latin, Averroes had by far the greatest impact on natural philosophy in the Latin West. Ironically, his Arabic works had almost no influence on Islamic thought, and his name is barely known in the Islamic world.

Some thirty-eight commentaries on the works of Aristotle are attributed to Averroes. This large number is explicable by the fact that Averroes wrote three different kinds of commentaries. These took the form of an epitome, or brief summary of Aristotle's text; a middle commentary, or paraphrase of Aristotle's text; and a long commentary, which was a detailed, sequential discussion of section after section of Aristotle's text. For some of Aristotle's treatises, Averroes wrote all three types of commentaries, while for others he wrote only two (middle and long). During the first half of the thirteenth century, Michael Scot and others translated fifteen of these thirty-eight Arabic commentaries into Latin. Virtually all of Averroes' commentaries were also translated into Hebrew for Jewish scholars in Western Europe. During the sixteenth century, nineteen of those treatises were translated from Hebrew into Latin. Because of this, Averroes' influence became even greater in the sixteenth century, especially in Italy.

If the translations from Greek to Latin were likely to be more accurate than those from Arabic, one might inquire whether the translations into Latin from Arabic were reliable. Were they sufficiently reliable to enable scholars in the Middle Ages to arrive at a reasonable consensus about Aristotle's ideas and interpretations? The answer, it would seem, is in the affirmative. Professor Harry A. Wolfson, a master of the Greek, Latin, Arabic, and Hebrew languages, and an authority on Aristotle, medieval philosophy, and translations, assessed the level of understanding of Aristotle's thought attained by the sixteenth-century Jewish Aristotelian scholar, Hasdai Crescas:

Aristotle was unknown to Crescas in the original Greek. He was also unknown to him in the Arabic translations. He was known to him only through the Hebrew translations which were made from the Arabic. It would be, however, rash to conclude on the basis of this fact that his knowledge of Aristotle was hazy and vague and inaccurate, for, contrary to the prevalent opinion among students of the history of philosophy, the translations of Aristotle both in Arabic and in Hebrew have preserved to a remarkable degree not only clear cut analyses of the text of Aristotle's works but also the exact meaning of his terminology and forms of expression. (Wolfson 1929, 7)

Wolfson's remarks about translations into Arabic and Hebrew apply equally, and probably even more so, to translations of Aristotle's texts into Latin. Once Aristotle's treatises on logic and natural philosophy were adopted as the basic curriculum in the arts faculties of medieval universities in the course of the thirteenth century, they became the common property of students and teachers over the whole of Europe for almost four centuries. Students read similar Aristotelian texts at the Universities of Oxford and Paris, and the scholars who wrote commentaries or questions on those same texts confronted much the same issues, and understood one another's interpretations and analyses, because they shared a common heritage of Aristotle's works. We should, however, keep in mind, as we have already seen that all the manuscripts of Aristotle's works were hand copied, making it inevitable that the different versions varied with small to large discrepancies. In general, however, medieval natural philosophers seem to have agreed on the meanings and interpretations of Aristotle's texts, although there were, of course, disagreements, as there still are today.

UNIVERSITIES IN THE MIDDLE AGES

As the Middle Ages entered the thirteenth century, two great events intermingled to produce a monumental change in the intellectual life of Western Europe: Universities were already in operation by 1200 at Oxford, Paris, and Bologna, and the translations of many of Aristotle's books on natural philosophy and logic were available and ready to serve as the basis of a new curriculum in the emerging universities. As evidence of this new relationship, as well as an early indication of the impact of the church on Aristotle's natural philosophy, one may point to a decree in 1210 by the Parisian Synod, which declared, among other things, that "no lectures are to be held in Paris either publicly or privately using Aristotle's books on natural philosophy or the commentaries, and we forbid all this under pain of excommunication" (Maccagnolo 1988, 429–430). The prohibition of 1210 is a good indication that Aristotle's treatises on natural philosophy, and perhaps also the commentaries by Averroes, were readily available, for otherwise there would have been no need to ban public and private lectures on their works. Although the University of Paris is not specifically mentioned in the decree, it is virtually certain that mention of public lectures is a reference to lectures at that university.

Although Aristotle's natural philosophy entered the University of Paris amidst controversy, the adverse reaction to his teachings was largely confined to that university and had little impact on the Universities of Oxford and Bologna. During the course of the thirteenth century, as the universities absorbed Aristotle's natural philosophy, serious tensions appeared from time to time between the arts and theological faculties of the University of Paris. Before describing these, however, it will be useful to summarize the essential features of the medieval university.

The university as we know it today has evolved from its first beginnings in the Middle Ages, when, as previously mentioned, it was already in existence in three different places by 1200. In short, the university is a medieval invention. Nothing like it had ever been developed in any preceding or contemporary civilization. The university did not appear on the scene full blown, but had an evolutionary history that begins with cathedral schools, a class of schools sponsored by cathedrals that flourished in Europe during the eleventh and twelfth centuries. Cathedral schools were attended by students who came from all over Europe, often attracted by a famous teacher or by the reputation of the school. Students in Europe were prepared to travel and move from one school to another to study with this or that teacher or group of teachers. As universities emerged, they attracted students from all over Europe, who flocked to university cities, just as students had earlier clustered into cathedral schools. No university could be founded without a congregation of teachers and students in relatively large numbers, a set of conditions that apparently materialized in cities like Paris, Oxford, and Bologna. Why these schools did not materialize in larger cities, such as London or Rome, is simply unknown (Rüegg 1992, 14).

Students and teachers were usually foreigners in the cities in which they studied and taught, and they therefore lacked the rights and privileges granted to the citizens of those communities. To strengthen their position in the communities and acquire rights and privileges to improve their situation as much as possible, the masters and students recognized the need for organization. To achieve their goals and enhance their prospects, they formed themselves into a corporation, or

Figure 6.1. A classroom in fourteenth-century Germany. (Staatliche Museen, Berlin.)

universitas, as had been done by numerous craft and merchant guilds (see Figure 6.1).

The corporation was an important medieval legal concept by means of which church and state in the Middle Ages granted individuals in the same profession or trade the right to form a fictional entity, which afforded its members various legal rights. For example, the members of the corporation had the right to elect their own officers. The candidate with the majority of votes was regarded as an elected officer. The corporation was thus a form of representative government. Elected officials had the right to represent the corporation in law courts, or before state and church authorities. Corporations could own property, sue or be sued in the courts, draw up contracts, and perform other functions on behalf of its members. To look out for the interests of their members, corporations formulated laws and statutes that were considered binding for their members.

Corporate status was a boon to universities. Such privileges effec-

tively conferred autonomy on all entities legally regarded as corporations. As educational institutions, the universities enjoyed autonomy
and were thus able to control their own affairs. As members of the
university community, students and teachers enjoyed various privileges and were also expected to adhere to certain restrictions. Despite
their legal autonomy, universities in the thirteenth and fourteenth centuries were subject to some ecclesiastical influence and pressure, but
they were also recipients of ecclesiastical benefits.

The corporate structure of the medieval university provided substantial stability and thus allowed the teaching of natural philosophy
to develop as the basis of all university learning in the four faculties
that comprised a major university, namely arts, theology, medicine,
and law (see Figure 6.2). The last three faculties constituted graduate
level instruction for professional degrees in theology, medicine, and
law. To matriculate for a degree in one of these higher faculties, the
prospective student was ordinarily expected to have acquired a Bachelor of Arts degree as well as a Master of Arts degree, the latter customarily requiring two additional years of lectures and study beyond
the four years required for the bachelor's degree. It was while pursuing their arts degrees that future theologians, physicians, and lawyers
acquired an unusual degree of familiarity with logic and natural philosophy. For approximately four centuries, the arts faculty gave all
university students a common intellectual experience and a common
intellectual heritage. Whatever their future careers might be—whether
as professors in arts, as practitioners or professors of theology, law, or
medicine, or as clerks in royal courts or municipalities—all students
acquired a basic knowledge of logic and natural philosophy, subjects
that were valued for their own sakes and also regarded as useful, if
not indispensable, in the higher disciplines of theology, law, and medicine.

TYPES OF LITERATURE IN NATURAL PHILOSOPHY

Medieval natural philosophers developed three basic formats for
presenting their subject. Two of the three formats were directly related
to Aristotle's works in natural philosophy. The third was concerned
with topics about subject themes in natural philosophy that were independent of Aristotle's natural philosophy, although Aristotle may
have briefly discussed some of those themes. For example, lengthy
treatises were composed on such subjects as whether the celestial mo-

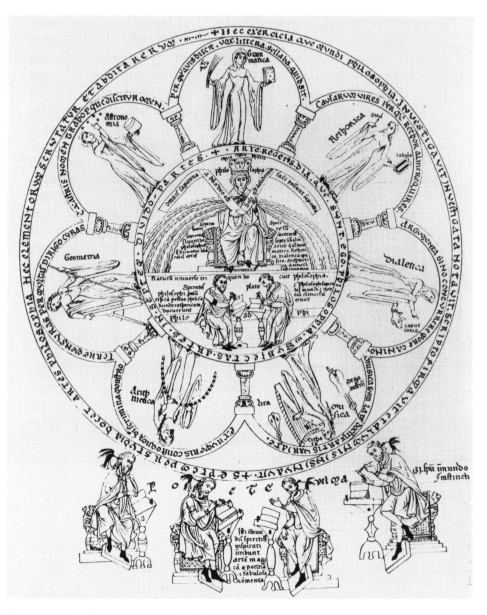

Figure 6.2. Philosophy surrounded by the seven liberal arts (arithmetic, geometry, music, astronomy, grammar, rhetoric, and dialectic, or logic), which are its offspring. A twelfth-century illustration from a collection made by the abbess Herrad of Landsberg (d. A.D. 1195). Although the collection was destroyed in 1870, the illustration shown here survived and has been reproduced by Rosalie Green, *Hortus Deliciarum*, 2 vols. (1979). (For more, see Murdoch 1984, 192, 381.)

tions were commensurable or incommensurable, on problems of projectile motion and the proper ways to represent those motions mathematically, and also on the ways to mathematically represent variations in the intensities of qualities. Some highly imaginative and significant contributions to natural philosophy were made in the genre of tractates.

The outstanding author in this genre of literature was undoubtedly Nicole Oresme, who was by far the most original and significant contributor on a range of topics. Indeed, he wrote on the first and third subjects just mentioned, and also composed an original treatise on atmospheric refraction titled *On Seeing the Stars* (*De visione stellarum*) (Oresme 2000). In this unusual work, Oresme makes a spectacular contribution to the history of science. He argues that refraction of light does not require a single refracting interface between two media of differing densities, as had been previously thought, but that light will be refracted along a curved path when it is in a single medium of uniformly varying density. For example, if air increases in rarity the farther its distance from the earth, light will pass through it along a curved path. To deduce a curved path, Oresme used his knowledge of convergent infinite series, assuming that successive refractions produced successive line segments and that as the line segments increased to infinity they form a curved line (Oresme 2000, 40–55). Danny Burton, who discovered these ideas in Oresme's work, observed that Robert Hooke and Isaac Newton were previously thought to have first argued that light is continuously refracted as it moves along a curved path through a uniformly decreasing medium. "While the definitive demonstration of the curvature of light in the atmosphere was Hooke's and Newton's," Burton explains, "the original argument for such curvature was Oresme's" (Oresme 2000, 53). Important as the lengthy tractates were, they do not convey the sense of Aristotelian natural philosophy as it was taught, studied, and written about in the universities for three or four centuries. For this, we must turn to the first two formats of literary production in natural philosophy.

The first of these was the commentary, the most important form of which was a section-by-section exposition of a treatise by Aristotle. The author would sometimes include a section of text and then explain it; this would be followed by the next section, which the author would also explain. The entire treatise was presented in this manner. Aristotle's full text was explained in brief successive segments. Sometimes in lieu of the full text, only the first few words of each section were presented, after which the author would discuss that section and then present the next few words of the next section, and so on through the entire treatise. The few words of each section were simply guide words to cue the reader into the section under discussion. Occasionally, paraphrases were made of Aristotle's text, or Aristotle's text and the author's opinions were intermingled.

By far the most important type of literature in medieval natural philosophy was the second format, which took the form of questions. The questions treatise was an outgrowth of oral disputations that were held by students and teachers in the classroom. In these disputations, the teaching master proposed a question for his class. Students were chosen to defend the affirmative and negative sides. After the presentation of both sides, the master was expected to resolve, or "determine," the question by proposing a solution. This became the skeletal frame of all questions that were included in questions treatises, or *questiones* as they were called in Latin. The structure of the typical written question was patterned after the following six-step outline:

1. Statement of the question, which usually began with the word "whether" (*utrum* in Latin). For example, the author asked: "whether the earth is spherical"; "whether there are four elements, no more nor less"; or "whether lightning is fire descending from a cloud," and so on.

2. Arguments opposed to the author's position, usually referred to as the "principal arguments" (*rationes principales*).

3. Assertion of one or more opinions opposed to the principal arguments, often accompanied by an appeal to a major authority, usually Aristotle.

4. Clarification about the meaning of the question or any of its terms; an optional step.

5. Author's main arguments, which were presented in a variety of ways. Sometimes the arguments were given as ordinally numbered conclusions (*conclusiones*); other times they were not identified as conclusions, but were numbered ordinally; or they were left unnumbered, but presented one after the other.

6. Brief refutation of each of the principal arguments presented in the second step.

Medieval natural philosophy was made up of hundreds of questions similar to the three included in the first step. Most were straightforward, based on Aristotle's descriptions of the structure and operations of the physical world. A considerable number, however, were about physical conditions and circumstances that were purely hypothetical and impossible in Aristotle's conception of the world. Although Aristotle had regarded them as impossible and absurd, medieval natural philosophers frequently asked whether God could create the conditions that Aristotle had deemed impossible, as, for example, whether it is possible that other worlds might exist, whether God could make a vacuum, or whether God could make accidents exist that did not inhere in any substance. The circumstances that produced such questions must now be described.

THE RELATIONS BETWEEN NATURAL PHILOSOPHY AND THEOLOGY IN THE THIRTEENTH CENTURY

As we saw, public and private lectures on Aristotle's works on natural philosophy were banned in 1210 in Paris, and therefore at the University of Paris, on pain of excommunication. This was the most drastic action taken by church authorities against natural philosophy during the thirteenth century, although a number of other noteworthy actions were taken in the latter half of that century. The ban was reissued in 1215, but it may well have been recognized as ineffective, because Pope Gregory IX modified it in 1231 when he issued a famous bull, or proclamation, known in its Latin form as *Parens scientiarum*. Although Gregory's proclamation is usually regarded as the Magna Carta of the University of Paris, it also included a different approach to the study of Aristotle at that same university. Instead of banning Aristotle's works, and losing all that was of value in those treatises, Gregory decided to purge them of all material that was offensive to the Catholic faith. To achieve this end, he appointed a three-man committee. As far as is known, the committee failed to perform its task, because no report has ever been found. It is safe to assume that Aristotle's works were not expurgated, and it is likely that the ban on lecturing remained in effect. Indeed, it seems to have remained in effect for approximately forty years, until around 1255. The ban did not apply to Aristotle's logical works or his treatises on ethics and politics. Despite the ban on public and private readings or lectures, it is quite likely that Aristotle's works on natural philosophy were read privately by interested students and teachers. Although the ban was even extended to the University of Toulouse in 1229, it was never imposed on the Universities of Oxford or Bologna.

By 1255, the ban on Aristotle's natural philosophy had apparently ended, as is obvious from a list of texts for that year revealing that all of Aristotle's treatises were used in lecture courses at the University of Paris. Even prior to 1255, Roger Bacon had been giving public lectures on Aristotle's natural books at the University of Paris, perhaps as early as the mid-1240s. Aristotle's books would never again be banned in Europe during the rest of the Middle Ages. Indeed, they were warmly welcomed. Nevertheless, more difficulties lay ahead, which now took another form. Instead of banning Aristotle's texts outright, or attempting to eliminate offensive material from those texts, theological authorities instead chose to condemn certain ideas that they regarded as dangerous or threatening to the faith. During the 1260s and 1270s, conservative theologians, among whom the most eminent was Saint Bonaventure (John Fidanza) (A.D. 1217–1274), launched a second major assault in what ultimately proved a vain effort to curtail the influence of Aristotle.

A sense of frustration was already apparent by 1270, when traditional theologians prevailed upon the bishop of Paris, Stephen (Etienne) Tempier, to condemn thirteen articles that were derived from the writings of Aristotle and Averroes, an act triggered by the activities of two famous arts masters in the 1260s, Siger of Brabant (c. A.D. 1240–d. c. 1284) and Boethius of Dacia (d. c. A.D. 1284). Siger and Boethius were dedicated followers of Aristotle and his Islamic commentator Averroes. They adopted some of the more extreme Aristotelian interpretations of Averroes, which included, among other ideas, "that the world is eternal, that terrestrial events are subject to the movement of the heavenly bodies, that there is only one human intellect, that there is no free will or choice, and that the soul is destroyed with the body and is unable to suffer punishment" (Asztalos 1992, 424). All of these views were contrary to Christianity and were included among the thirteen articles condemned by Bishop Tempier in 1270.

Although, as we shall see, the condemnation of the thirteen articles seems to have had little effect, the teaching masters in the arts faculty sought to avoid possibly dangerous entanglements with theology by demarcating the boundaries between natural philosophy and theology. They did this in 1272 by making it mandatory for all members of the arts faculty to swear an oath in which each agreed not to consider theological questions—as for example, the Trinity and incarnation—in their treatises on natural philosophy. If for some reason an arts master was unable to avoid a theological problem, he was sworn to re-solve the issue in favor of the faith. In general, however, arts masters were expected to avoid the citation of theological arguments in their treatises on natural philosophy.

The Condemnation of 1270 and the adoption of an oath by members of the arts faculty in 1272 were apparently insufficient to placate the conservative theologians. Sometime between 1268 and 1274, Giles of Rome (c. A.D. 1243–1316), a member of the Augustinian Order and the first member of his order to become a master of theology at the University of Paris, wrote a treatise titled *Errors of the Philosophers*. Giles' objective is immediately apparent in the opening words of the treatise: "Here begin the errors of the philosophers Aristotle, Averroes, Avicenna, Algazel, Alkindi and Maimonides Collected by Brother Giles of the Order of St. Augustine" (Giles of Rome 1944, 3). The "errors" were those by Aristotle and four of his Islamic followers, Averroes (c. A.D. 1126–1198), Avicenna (A.D. 980–1037), Algazel (A.D. 1058–1111) and Alkindi (d. c. A.D. 870), and one Jewish follower, Maimonides (A.D. 1138–1204). (As was usual in the Latin Middle Ages, Giles mistakenly included Algazel with the followers of Aristotle. In truth, the opinions of Avicenna were improperly attributed to Algazel, who was then assumed to be a follower of Avicenna; Algazel was actually hostile to Aristotle and his followers.) Giles identifies fourteen errors for Aristotle, twelve errors for Averroes, twenty-two errors for Avicenna, twelve errors for Algazel (more accurately Al-Ghazali, but known as Algazel to Latin scholars), fifteen errors for Alkindi, and fifteen errors for Maimonides. Some of the same errors appear for more than one author. Many of them concern some aspect of the eternity of the world. For example, eight of Aristotle's fourteen errors concern eternity, as we see:

1. That motion did not have a beginning.

2. That time is eternal.

3. That the world did not have a beginning.

4. That the heavens were not made.

5. That God could not make another world.

6. That generation and corruption had no beginning and will have no end.

7. That the sun will always cause generation and corruption in the sublunary world.

12. That it is impossible to admit a first man or a first rainfall. (Giles of Rome 1944, 11–13)

In addition to the fifth error, in which Aristotle denied God the power to create another world, Giles of Rome denounces Aristotle for other restrictions on the divine power, as we see in the following errors:

8. That nothing new can proceed directly from God.
9. That the resurrection of the dead is impossible.
10. That God could not make an accident without a subject.
13. That two bodies cannot in any way be in the same place. (Giles of Rome 1944, 13)

The other philosophers repeat some of Aristotle's errors and add many of their own. A fair number of the errors Giles detected in the six non-Christian philosophers would reappear in the great Condemnation of 1277. Before turning to that condemnation, however, it will be helpful to describe the antagonisms that had developed between the faculties of arts and theology at the University of Paris by the 1270s, a development that played a significant role in generating the Condemnation of 1277.

Tensions Between the Faculties of Arts and Theology

As separate corporate entities, the faculties of arts and theology were each the guardian of a significant discipline: the arts faculty overseeing natural philosophy and the theology faculty defending and interpreting matters of faith and revelation. In a society in which the Christian religion played a dominant role, it is easy to see that, solely by virtue of their subject matters, the theology faculty had far greater power and influence than did the arts faculty. Many, if not most, thirteenth-century theologians regarded revelation as superior to all forms of knowledge. It is, therefore, hardly surprising that, in general, theologians retained the traditional opinion that secular learning was the handmaiden of theology, and consequently that science and natural philosophy existed for the benefit of theology. As a firm believer in this doctrine, Saint Bonaventure wrote in support of this attitude (see Figure 6.3). In a treatise titled *Retracing the Arts to Theology* (*De reductione artium ad theologiam*), Bonaventure upheld the superiority of theology as the "queen of the sciences" (on theology as a science, see below), a judgment based on the firm conviction that all knowledge is derived by divine illumination from sacred scripture. Theologians

Figure 6.3. Saint Bonaventure. From a fresco painted by Fra Angelico around 1450 and now in the Vatican.

like Bonaventure saw no conflict between faith and reason, provided one understood that the former guides the latter.

By contrast, masters in the faculties of arts viewed themselves as the custodians of philosophy and therefore of reasoned analysis. For them, natural philosophy was the key to understanding the physical cosmos. They certainly did not regard it as a handmaiden to theology, but as a subject to be studied for its own sake, because it was the only way to understand the workings of the physical world. Although they disagreed often enough with Aristotle's ideas and interpretations, arts masters had the highest regard for Aristotle's works and his use of reasoned argument. They successfully made natural philosophy autonomous from theology, a process that will be described later. In the 1270s, however, tensions remained high.

The Condemnation of 1277

The *Errors of the Philosophers* compiled by Giles of Rome was but a sign of the theological and intellectual turmoil that gripped Paris in the 1260s and early 1270s. As a theologian and member of the Augustinian Order, Giles was greatly disturbed by the large body of literature that espoused ideas judged contrary to the Christian faith. He was not alone. On January 18, 1277, Pope John XXI notified Stephen Tempier that he had heard reports of heretical opinions in the Paris area. He asked to be informed of the situation. It has been conjectured, on plausible grounds, that Bishop Tempier was already investigating heretical opinions at the University of Paris by the time he received the papal letter. On March 7, 1277, Bishop Tempier, acting on the advice of his theological consultants, issued a sweeping condemnation of 219 propositions. To uphold or defend any one of these propositions, or articles, was to do so at pain of excommunication. (In England, on March 18 of the same year, Robert Kilwardby, archbishop of Canterbury, condemned 30 similar propositions that applied to Oxford University.)

Certain of the Paris articles reveal the deep antagonism that had been engendered between arts masters and theologians, as is apparent from the following six articles:

37. That nothing should be believed unless it is self-evident or could be asserted from things that are self-evident.
40. There is no higher life than philosophical life.

152. That theological discussions are based on fables.

153. That nothing is known better because of knowing theology.

154. That the only wise men of the world are philosophers.

175. Christian Revelation is an obstacle to learning. (Gilson 1938, 64; Grant 1974, 48–50)

By including these articles, the theologians seem to have exacted a measure of revenge for alleged slights against them and their discipline. The 219 articles, however, were apparently drawn up hastily and with little regard for any proper order. Although the articles were distinct, many repeated the same error in different guises—especially with respect to the eternity of the world, of which there seem to be approximately twenty-seven different articles. Almost anything that occurred to the members of the selection committee seems to have been judged worthy of inclusion. Indeed, some, if not many, of the condemned articles were probably not even drawn from written texts, but may have been derived from conversations involving teachers or students in a classroom or dormitory, or simply from hearsay.

Many, if not most, of the condemned articles were in some way drawn from Aristotle's natural philosophy. Two of the most important themes relevant to that natural philosophy involved articles about the eternity of the world, and limitations on God's absolute power to do anything he pleases, short of a logical contradiction. Because, as mentioned earlier, approximately twenty-seven articles concern the eternity of the world, only some of the more striking ones will be mentioned here:

9. That there was no first man, nor will there be a last; on the contrary, there always was and always will be generation of man from man.

87. That the world is eternal as to all the species contained in it; and that time is eternal, as are motion, matter, agent, and recipient; and because the world is [derived] from the infinite power of God, it is impossible that there be novelty in an effect without novelty in the cause.

91. That the argument of the Philosopher [i.e., Aristotle] demonstrating that the motion of the sky is eternal is not sophistical; and it is amazing that profound men do not see this.

93. That celestial bodies have eternity of substance but not eternity of motion.

101. That an infinite [number] of celestial revolutions have preceded which it was not impossible for the first cause [that is, God] to comprehend, but [which are impossible of comprehension] by a created intellect.

107. That elements are eternal. However, they have been made [or created] anew in the relationship which they now have.

202. That the elements have been made in a previous generation from chaos; but they are eternal. (Grant 1974, 48–50)

Condemned articles concerned with limitations on God's power to do as he pleases short of a logical contradiction are of great importance for the relations between science and religion in the Middle Ages—perhaps of greater significance than those about the eternity of the world. Articles relevant to this theme included the following:

34. That the first cause [that is, God] could not make several worlds.

35. That without a proper agent, as a father and a man, a man could not be made by God [alone].

38. That God could not have made prime matter without the mediation of a celestial body.

48. That God cannot be the cause of a new act [or thing], nor can He produce something anew.

49. That God could not move the heavens [or world] with a rectilinear motion; and the reason is that a vacuum would remain.

63. That God cannot produce the effect of a secondary cause without the secondary cause itself.

141. That God cannot make an accident exist without a subject nor make several dimensions exist simultaneously.

147. That the absolutely impossible cannot be done by God or another agent—An error, if impossible is understood according to nature. (Grant 1974, 48–50; Article 63 is from Lerner and Mahdi 1963, 343)

There were, of course many other condemned propositions, some of which were concerned with physical and cosmological problems involving orbs, planets, and elements. There were also attacks against the tendency in the arts faculty to rely heavily on reason and reasoned argumentation. Thus, Article 90 condemned the idea that there could be two truths, one in natural philosophy, the other within the faith. It denounced the idea "that a natural philosopher ought to deny absolutely the newness [that is, the creation] of the world because he depends on natural causes and natural reasons. The faithful, however, can deny the eternity of the world because they depend on supernatural causes" (Grant 1974, 48). Scholars have yet to discover anyone in the Middle Ages who subscribed to the "doctrine of the double truth"

(Grant 1996, 76–78), even though Bishop Tempier, in a message introducing the 219 propositions speaks as if it were a common practice. Indeed, one gets the impression that Tempier's rationale for issuing the condemnation was instigated by the rampant use of the doctrine of the double truth, as seems apparent from the letter he sent along with his condemnation. In this letter, Bishop Tempier denounced those who "say that things are true according to philosophy but not according to the Catholic faith; as if there could be two contrary truths, and as if contrary to the truth of Sacred Scripture there could be truth in the statements of the damned gentiles of whom it was written: 'I will destroy the wisdom of the wise' [1 Cor. 1:19] because true wisdom destroys false wisdom" (Grant 1974, 47).

The bishop and his theological advisors were upset by the boldness with which natural philosophers were using reason to analyze all sorts of questions. The natural philosophers seemed to have set no limits to their spirit of inquiry, as they sought to subject all knowledge and ideas to philosophical and scientific analysis, an attitude the bishop of Paris condemned in the following article:

145. That no question is disputable by reason which a philosopher ought not to dispute and determine, since arguments are taken from [or based on] things. Moreover, philosophy has to consider all things according to its diverse parts.

IS THEOLOGY A SCIENCE?

Although I shall later explain how certain condemned articles impacted natural philosophy in the fourteenth century, it is important now to consider further the relations between theology and natural philosophy. Even as the tensions between the arts masters and the theologians played a significant role in the thirteenth century, another aspect of their relationship centered around Peter Lombard's *Sentences*, the great twelfth-century theological textbook that was second only to the Bible in popularity (see chapter 5 for more on the *Sentences*). In the prologue to his book, Peter Lombard discussed the epistemic nature of theology, but it remained for Alexander of Hales, who was the first commentator on the *Sentences*, to consider whether theology is a science. Thereafter, all commentators on the *Sentences* felt an obligation to consider the status of theology: was it, or wasn't it, a science? In his *Commentary on the Sentences*, Thomas Aquinas laid the basis for an approach to theology that most theologians accepted during the Middle Ages (see Figure 6.4). He presented an argument in favor of theology as a science that won widespread acceptance. Thomas' most mature formulation of these arguments occurs in his great *Summa of Theology* (*Summa theologiae*, or, as it is more commonly called, *Summa theologica*). In the very first question of the treatise (Part I, question 1), Thomas considers "The Nature and Domain of Sacred Doctrine" in ten articles (Thomas Aquinas 1948, 3–19). In the second article, he asks "whether sacred doctrine [that is, theology] is a science?" Thomas explains that sacred doctrine is a science, but, he continues,

We must bear in mind that there are two kinds of sciences. There are some which proceed from principles known by the natural light of the intellect, such as arithmetic and geometry and the like. There are also some which proceed from principles known by the light of a higher science: thus the science of optics proceeds from principles established by geometry, and music from principles established by arithmetic. So it is that sacred doctrine is a science because it proceeds from principles made known by the light of a higher science, namely the science of God and the blessed. Hence, just as music accepts on authority the principles taught by the arithmetician, so sacred science accepts the principles revealed by God. (Thomas Aquinas 1948, 5–6)

Thomas goes on to argue that theology, or sacred doctrine, is a single science (Article 3), and that it is nobler and more transcendent than all other speculative and practical sciences (Article 5). In explaining his claim in the fifth article, Thomas declares that

one speculative science is said to be nobler than another either by reason of its greater certitude, or by reason of the higher dignity of its subject-matter. In both these respects this science surpasses other speculative sciences: in point of greater certitude, because other sciences derive their certitude from the natural light of human reason, which can err, whereas this derives its certitude from the light of the divine knowledge, which cannot err; in point of the higher dignity of its subject-matter, because this science treats chiefly of those things which by their sublimity transcend human reason, while other sciences consider only those things which are within reason's grasp. (Thomas Aquinas 1948, 9)

Because it derives its certitude directly from God, and because it treats things that transcend human reason, theology is superior in every basic way to all other sciences. Thus did Saint Thomas make

Figure 6.4. Saint Thomas Aquinas. From a fifteenth-century painting by Justus of Ghent, which was based on an earlier copy.

theology a science that was more powerful and important than the exact sciences, such as astronomy and optics, and the mathematical sciences of arithmetic and geometry. No secular science could surpass theology, because the latter derived its first principles from God and sacred scripture. There was no arguing with that claim. Theology was therefore superior to natural philosophy and logic, sciences that were routinely used in theology.

But the converse was not true: natural philosophy did not use theology when it was treating natural phenomena. For one thing, by the Oath of 1272, arts masters swore they would not deliberately involve theology in their discussions. Indeed, although natural philosophy was regarded by the theologians as less important than theology, and even viewed as a handmaiden to theology, that was no longer true for natural philosophers. When Thomas Aquinas made theology an independent science, distinct from all others, he inadvertently conferred autonomy on natural philosophy. Although theologians regarded theology as the "queen of the sciences," arts masters could now view natural philosophy as a scientific discipline completely independent of theology. As a superior science, theology would not be mixed with an inferior science such as natural philosophy. Thus, the emergence of theology as an independent science in the second half of the thirteenth century had an inadvertent corollary: a guarantee that natural philosophy would also be regarded as an independent science. In the interrelations between the two sciences, as we shall see, it was theology that relied heavily on natural philosophy, whereas theology had no substantive role in natural philosophy.

Albertus Magnus, a theologian and an outstanding natural philosopher, as well as the teacher of Thomas Aquinas, provides early but significant evidence of this relationship (see Figure 6.5). Albertus' Dominican brothers had requested that he write a book about nature that would enable them to understand Aristotle's books in a proper manner. In the opening passage of his commentary on Aristotle's *Physics*, Albertus informs his fellow Dominicans that he will not speak about divine matters, because such matters "can in no way be known by means of arguments derived from nature." He goes on to explain that "we take what must be termed 'physics' more as what accords with the opinions of Peripatetics [i.e., Aristotelians] than as anything we might wish to introduce from our own knowledge . . . for if, perchance, we should have any opinion of our own, this would be proffered by us (God willing) in theological works rather than in those on

Figure 6.5. Albertus Magnus (Albert the Great). Fresco by Tommaso da Modena (1352), located at the Monastery of San Niccolò, Treviso. Alinari/Art Resource N.Y. (see Lindberg 1992, 228).

physics" (Synan 1980, 10). In his *Commentary on Aristotle's On the Heavens*, Albertus reveals the same desire to keep natural philosophy and theology separate, when, in a discussion inquiring whether heaven is ungenerable and incorruptible, he explains:

Another opinion was that of Plato who says that the heaven was derived from the first cause by creation from nothing, and this opinion is also the opinion of the three laws, namely of the Jews, Christians, and Saracens. And thus they say that the heaven is generated, but not from something. But with regard to this opinion, it is not relevant for us to treat it here. (Grant 2001, 192)

Although Albertus was wrong to attribute to Plato a belief in a creation from nothing, it is noteworthy that he refuses to include a discussion of the doctrine of creation from nothing in a straightforward treatise on natural philosophy. Creation from nothing was a divine, supernatural act and therefore not a legitimate topic for a book on natural philosophy. Thus did Albertus Magnus separate the sciences of physics, or natural philosophy, from theology. His pupil, Saint Thomas Aquinas, would have undoubtedly agreed with the master's sentiments.

The thirteenth century laid a foundation for the interrelations between science and religion in the fourteenth and fifteenth centuries. That century was the first in which Christianity came face to face, so to speak, with Aristotle's extensive and powerful secular natural philosophy. The relations between the two—in effect, the relations between science and religion, or natural philosophy and theology—were often acrimonious and bitter. Theology and the power of the church were sufficient to curb and limit the ambitions of the arts masters, who sought, as much as possible, to give free reign to their efforts to interpret the physical cosmos in straightforward Aristotelian terms, unencumbered by theological restrictions and limitations. As the dust settled in the fourteenth century, it became obvious that theologians had an enormous degree of latitude to use natural philosophy and analytical methods virtually as they pleased in their theological treatises, as is evident in their *Commentaries on the Sentences of Peter Lombard*. By contrast, arts masters usually sought to avoid introducing theology into their commentaries and questions on the books of Aristotle's natural philosophy. Not only was it potentially dangerous for an arts master untrained in theology to intermingle natural philosophy with theology, but the Oath of 1272, which was in effect for much of the fourteenth century at the University of Paris, forbade arts masters from involvement with theological issues. In the next chapter, we shall see how theologians intermingled theology and natural philosophy, and how arts masters treated theological issues when these sometimes intruded into problems of natural philosophy.

Chapter 7

·•· ⇥◆⇤ ·•·

The Interrelations between Natural Philosophy and Theology in the Fourteenth and Fifteenth Centuries

The Oath of 1272 and the Condemnation of 1277 coalesced in a significant episode that involved John Buridan, a famous arts master and natural philosopher at the University of Paris. During the 1330s, Buridan had occasion to compose *Questions on the Eight Books of the Physics of Aristotle.* In Book 4, question 8, Buridan asks "whether it is possible that a vacuum exist by means of any power?" After presenting arguments denying the possibility of a vacuum, Buridan rejects this position by declaring:

The opposite [position] is argued because God could annihilate everything under the lunar orb with the magnitude and figure of the lunar orb preserved. Then the concave orb of the moon, which is now a plenum in the lower world, would be a vacuum, just as a pitcher would be a vacuum if God annihilated the wine in it while preserving the pitcher and where no other body enters or is made in the pitcher. And thus some of my lords and masters in theology have reproached me on this, [saying] that sometimes in my physical questions I intermix some theological matters which do not pertain to the artists [that is, Masters of Arts]. But with [all] humility I respond that I very much wish not to be restricted [with respect] to this, namely that all masters beginning in the arts swear that they will dispute no purely theological question, nor [dispute] on the incarnation; and they swear further that if it should happen that they dispute or determine some question which touches faith and philosophy, they will determine it in favor of the faith and they will destroy the arguments (*rationes*) as it will be seen that they must be destroyed.

Now it is evident that if any question touches faith and theology, this is one of them, namely whether it is possible that a vacuum exist. And so, if I wish to dispute it, it is necessary that I say about it what appears to me must be said according to theology, or to perjure myself and avoid the arguments on the opposite side insofar as this will seem possible for me. But I could not resolve these arguments [on the opposite side] unless I produce them. Therefore, I am compelled to do these things. I say, therefore, that "vacuum" can be imagined in two ways. (Grant 1974, 50–51)

The two ways in which Buridan imagines how a vacuum could exist are of great interest. The second way is the one mentioned above, namely that God could make a vacuum within the lunar sphere by virtue of his absolute power. That conception of a vacuum does not correspond to any particular article condemned in 1277, although Buridan's description is consistent with the emphasis on God's absolute power to perform any act that is naturally impossible in Aristotle's natural philosophy. Indeed, it is compatible with article 49, which denied that God could move the entire world with a rectilinear motion, because a vacuum would be left behind, thus implying that the impossibility of a vacuum would prevent God from moving the world. The implication is clear: God could not only move the world rectilinearly, but by virtue of its motion, a vacuum could be left behind, even though Aristotle had argued that the existence of a vacuum is impossible. In the first mode of conceiving a vacuum, Buridan explains:

Firstly, as to the first mode of imagining a vacuum, I assume that God can make an accident without a subject and is able to separate accidents from their subjects and to conserve them separately. And thus He is able to create an absolute dimension without there being any substance or accident distinct from it. Secondly, it seems to me that the penetration of dimensions is not impossible by God. Indeed, He is able to make several bodies exist simultaneously in the same subject, or in the same place, without their mutually differing from each other with respect to location, namely that one [body] should be outside the other with respect to location [or place]. Therefore, God can make an absolute dimension or space separated from every natural substance in which [dimension or space] natural bodies can be received. And this is called a vacuum according to the first way of speaking related previously. (author's translation from Buridan 1509, fol. 74r, col. 1)

This definition has strong connections with Article 141 of the Condemnation of 1277, which declares: "That God cannot make an accident exist without a subject, nor make several dimensions exist simultaneously [in the same place]" (slightly altered from Grant 1974, 49). In this first explanation of a vacuum, which Buridan does not seem to take seriously and does not discuss further, he defines a vacuum as a dimension that can receive other dimensions that exactly coincide with its own dimensions, or are smaller. Although Buridan does not specifically mention any articles from the Condemnation of 1277, he must, at the very least, have had Article 141 in mind.

But why did Buridan feel he had to apologize to the theologians? There is nothing in his question that violated either the Oath of 1272, or the Condemnation of 1277. In the questions cited above, Buridan concedes, however, "if any question touches faith and theology, this is one of them, namely whether it is possible that a vacuum exist." It was a theological question only in the sense that the vacuum was regarded as naturally impossible, but creatable by divine power. Acknowledging that it was a question that "touches faith and theology," Buridan justified inclusion of it in his *Questions on Aristotle's Physics* by appeal to the Oath of 1272. If he included a question touching faith and theology, the oath required that he resolve it in favor of the faith. In order to do this, however, he had to include the theological arguments, or perjure himself.

Sometime between 1506 and 1511, approximately 160 or 170 years after Buridan wrote his treatise, Luis Coronel (fl. A.D. 1511) wrote a commentary on Aristotle's *Physics* and cited Buridan's question about the possibility of a vacuum. After accurately recounting Buridan's discussion, Coronel declares: "These remarks of Buridan have astonished me, first, that our masters blamed him, for from the declaration of this term, *vacuum*, to conclude that it cannot be produced naturally, but can happen supernaturally seems in no way blameworthy" (Thorndike 1944, 87). No more than modern readers could Coronel discover why theologians would have objected to Buridan's discussion about vacuum.

Whatever the merits of Buridan's dispute with the theologians, he presented a succinct description of the oath and what was expected of arts masters in the fourteenth century. One can see the difficulties they faced. The Oath of 1272 did not absolutely forbid them from discussing theology, but it placed the burden of uncertainty on any arts masters who chose to enter the domain of theology. How far could they go? Was introducing a theological question worth the anxiety that might ensue? Could an arts master be confident that he had refuted an argument against the faith to the satisfaction of the theologians? Would a theologian find fault with his arguments and accuse him of heretical ideas and notions? Luis Coronel captures this uncertainty in his discussion of Buridan's question. In the third of five points, Coronel criticizes the Oath of 1272:

Thirdly, that oath does not seem reasonable when it compels a man to overthrow arguments, for there might be some loyal teacher for whom some argument worked out contrary to the determination of the church, and how would he overthrow it? But to this it seems it should be said that, if the instructor does not know how to overthrow such an argument, he ought not to formulate it in public to his students. But if he should do otherwise, he would remain perjured if he had taken the oath, and will sin although no oath has preceded. (Thorndike 1944, 87)

Then, drawing the seemingly inevitable inference from these difficulties, Coronel concludes his third point with this admonition: "Therefore, let Parisian teachers of artists who touch on theological problems look out for themselves."

By the end of the fifteenth century, the Oath of 1272 was apparently no longer required of arts masters, as Luis Coronel informs us in the fourth point about Buridan's travails: "I, inadequate and unworthy as I am, do not recall that when I was promoted to the degree in arts I took, or knew of any of my fellows taking, such an oath, but, alas, that laudable custom of the university along with others had become obsolete." Thus, despite his criticism of the oath, Coronel seems to favor an oath of some kind, as seems apparent from his closing remarks on Buridan when he declares: "And whether there was an oath or not, I would strive where occasion offered to proceed conformably to the intention of the person who ordered that oath be taken" (Thorndike 1944, 88).

Elsewhere in his *Questions on the Physics* and in other commentaries on Aristotle's works, Buridan was more self-effacing with respect to the theologians. In Book VIII, question 12 of *Questions on the Physics*, Buridan observes that "the Bible does not state that appropriate intelligences move the celestial bodies," so that "God, when He created the world, moved each of the celestial orbs as he pleased." Buridan suggests that instead of an external agent, such as intelligences, God may have chosen to impress a force, or impetus, into each celestial orb so that it could move by its own power. "And these impetuses which He impressed in the celestial bodies," Buridan explains, "were not decreased or corrupted afterwards, because there was no inclination of the celestial bodies for other movements. Nor was there resistance which would be corruptive or repressive of that impetus." Realizing that he was perhaps presumptuously suggesting the way God might have behaved when He created the world, Buridan then declares: "But this I do not say assertively, but [rather tentatively] so that I might seek from the theological masters what they might teach me in these matters as to how these things take place" (Clagett 1959, 536; Grant 1974, 277–278). In his *Questions on Aristotle's On the Heavens*, Buridan reacts in a similar manner. In Book 1, question 20, he declares: "Thirdly, I say that there is no body beyond the heaven or world, namely beyond the outermost heaven; and Aristotle assumes this as obvious. But you ought to have recourse to the theologians [in order to learn] what must be said about this according to the truth of faith or constancy" (Grant 1974, 51, n. 4).

Buridan's uneasiness and sense of uncertainty with the theologians was not a lasting phenomenon. Few were as diligent as Buridan, who chose to introduce theological considerations where he thought it necessary to respond properly to a question. Later in the century other major arts masters, such as Albert of Saxony, Marsilius of Inghen, and others took up similar questions, as well as questions that Buridan would probably not have considered. A description of some of these questions will enable readers to gain some insight into the kinds of hypothetical, counterfactual discussions that became commonplace in treatises on natural philosophy and theology during the latter part of the fourteenth century. Most of the questions involve reactions to articles condemned in 1277.

THE INFLUENCE OF THE CONDEMNATION OF 1277 ON NATURAL PHILOSOPHY

Numerous and almost unavoidable discussions on the possible existence of a vacuum, or void space, formed an important aspect of medieval natural philosophy. These discussions derived from Aristotle's rejection of the possibility of a vacuum in the fourth book of his *Physics*. Because the vacuum was a major topic of discussion for Aristotle, there would have been discussions about this phenomenon with or without a theological Condemnation in 1277. Not only would scholastic natural philosophers have discussed the possible existence of a vacuum, but on the hypothetical assumption that a vacuum did exist, they would have followed Aristotle's argument and inquired whether successive, finite motions could occur in a totally empty space. Medieval natural philosophers frequently used the phrase "nature abhors a vacuum" to show their basic agreement with Aristotle. But they departed from Aristotle's framework because of the Condemnation of 1277. Where Aristotle emphatically declared that a vacuum is impossible in nature, medieval natural philosophers and theologians, after the Condemnation of 1277, were obligated to assert that God could, if he wished, create a vacuum within or outside of the world. This becomes obvious in the context of Buridan's discussion cited above, in which Buridan argued that if God wished, he could create a vacuum within the concavity of the lunar region, despite Aristotle's denial of the possibility of a vacuum existing anywhere. In his *Questions on the Physics* (Book 4, question 11), Albert of Saxony considered whether a body could move in a vacuum that God had supernaturally created (see Grant 1974, 335, for a translation).

The Supernatural Possibility of Other Worlds

The most striking impact of the Condemnation of 1277 on natural philosophy involved Articles 34 and 49. Article 34, as cited in Chapter 6, declared: "That the first cause [that is, God] could not make several worlds" (Grant 1974, 48; brackets added). Aristotle had argued that nothing whatsoever could exist beyond our world, explicitly mentioning matter, place, vacuum, and time. It follows that no world could exist beyond our world, by which Aristotle meant that no world like our own could exist beyond ours. The existence of other worlds is impossible, Aristotle argued, because all the matter in existence is contained within our world, with none available to form additional worlds. Aristotle's arguments were already criticized in the ancient world. Some argued that matter could indeed exist beyond the world. In his *Commentary on Aristotle's On the Heavens,* the sixth-century Greek commentator Simplicius recounted a reaction to Aristotle's arguments by Stoic philosophers who repeated an even earlier counter-example that served to undermine Aristotle's position. If someone at the extremity of the world extended an arm beyond that extremity, what would happen? One of two alternatives, they concluded: Either the arm meets an obstacle and cannot be extended further; or, it meets no obstacle and can be extended beyond the world. In the first possibility, the Stoics imagined that the person would climb to the extremity of the obstacle and again extend an arm. The arm would either meet another material obstacle, or it would not. This can be repeated indefinitely, but in a finite world, the person will eventually reach the end of material obstacles and encounter void space. Because Simplicius' treatise was translated into Latin in the thirteenth century, the argument was known in the West and occasionally cited. But this argument in itself might have had only a slight effect and posed little threat to Aristotle's basic interpretation that the world is a finite, spherical cosmos with nothing whatsoever lying beyond it.

With the appearance of Article 34, all this changed. After 1277, it became mandatory for students and teachers at the University of Paris to concede that by virtue of his absolute power, God could create as many worlds as he pleased beyond ours (see Figure 7.1). Although there is no evidence of anyone who believed that God had in fact created other worlds in addition to ours, the possibility that he could was significant because it meant that contrary to Aristotle's claims, the existence of other worlds was not impossible, but, by virtue of God's absolute power, quite possible. The possibilities were intriguing and prompted natural philosophers to raise two basic questions: (1) whether any thing (space, void, place, time, or matter) could exist beyond the boundaries of our world; and (2) whether other worlds exist or could exist (for a list of discussants, see Grant 1994, 689–692). On the assumption that God could create as many worlds as he pleased, questions were raised about the manner in which he might do this: Would he make concentric or eccentric worlds—that is, worlds within worlds? Would he make one world after another in an unending succession of worlds? Or would he make a multiplicity of worlds simultaneously? Most of the discussions about a plurality of worlds focused on simultaneously created worlds, all of which were assumed to operate by the same laws as our own.

The surprising result was that most scholastic natural philosophers concluded that each world was a self-contained system with its own center and circumference. The same physical laws were assumed valid in each world. The behavior of elements and bodies in one world would not, and could not, affect the elements and bodies of another world. No element from one world would have any inclination to join its elemental counterparts in another world. Thus did medieval natural philosophers abandon Aristotle's basic idea that only one center and circumference could exist. They assumed that a multiplicity of equal centers and circumferences could exist simultaneously, one pair

Figure 7.1. God holding the world he created; from a French *Bible moralisée* of around 1250. (MS Bodley 270b, fol. 1r in Bodleian Library, Oxford, England.)

for each world. Thus, the elements of each world moved towards or away from their own center and circumference without any inclination to move toward the center or circumference of any other world.

Despite a strong sense that, contrary to Aristotle, the existence of a plurality of worlds was an intelligible concept, and that God could make as many other worlds as he pleased, no one believed that God had actually done so, or would ever do so. It was, nevertheless, a momentous departure from Aristotle's concept of the world. What Aristotle had regarded as naturally impossible was viewed in the Middle Ages as supernaturally possible. The existence of other worlds was not an absurdity, as Aristotle had argued, but an intelligible possibility, albeit only by divine command.

The Existence of Void Space beyond the World

The possible existence of other worlds was also assumed to involve the inevitable existence of empty spaces between adjoining worlds. At best, spherical worlds could meet and touch at only one point. Between any other points on their spherical surfaces there would either be more matter or void space. The idea that matter might lie between them was quickly dismissed, because the matter would belong to no particular world and therefore have no rationale for its existence. There was, however, general agreement that if other worlds were created, void spaces would exist between them.

While all arts masters and theologians were agreed that God could create vacuums beyond the world if he wished, medieval discussions about the existence of void space largely ignored questions about whether God could create a finite or infinite space beyond the world. Instead, they linked questions about God's location with his omnipresence, which was always assumed to be in an infinite void space. Because of God's infinite omnipresence, questions about extra-cosmic void space—almost always infinite extracosmic void space—fell to the theologians, who were the only ones thought qualified to discourse on such a subject. Arts masters did not dare offer opinions on God's location and the manner in which he is present in the world.

Along with whether God could create other worlds, Article 49 of the Condemnation of 1277 also generated hypothetical arguments that produced important ideas about how things might differ from Aristotle's view of the world. As cited earlier, Article 49 declared: "God could not move the heavens [or world] with a rectilinear motion; and the reason is that a vacuum would remain" (Grant 1974, 48; slightly altered). One had to concede that God could indeed move the world with a rectilinear motion, whether or not a vacuum was left behind. If God moved the world with a rectilinear motion, three fundamental Aristotelian principles would be violated: (1) the movement of the world would leave behind a void place, which is impossible in Aristotle's world; (2) a rectilinear motion of the world would not be classifiable as any of the three natural motions that Aristotle had distinguished, namely rectilinear up or down motions, and circular motion; and (3) it would violate Aristotle's assumption that motion is always from one material place to another material place—that is, that motion can occur only in a material plenum. But Aristotle had argued that no matter exists beyond the world and hence no places, and therefore, no motions are possible. Consequently, if God moved the world rectilinearly, its motion would be independent of places, a consequence that Aristotle would have regarded as absurd, but certain medieval natural philosophers regarded as plausible. In the fourteenth century, Nicole Oresme, for example, regarded a rectilinear motion of the entire world as an absolute motion independent of places. An absolute motion could not be related to any other body because, for the purposes of the argument, no other bodies existed outside of our world. Despite Aristotle's opinions, Oresme regarded such an absolute rectilinear motion as plausible and intelligible. Indeed, in his famous controversy with Gottfried Leibniz (A.D. 1646–1716), Samuel Clarke (A.D. 1675–1729), Isaac Newton's spokesperson, also regarded such a rectilinear motion as an intelligible concept.

The numerous discussions about God and extra-cosmic space are not just relevant to the relationship of science and religion, but are vital to straightforward history of science, because medieval arguments on infinite void space influenced some of the great scientific thinkers of the seventeenth century, including the greatest, Sir Isaac Newton. The theme of an infinite void space beyond our world became one area of discussion where quite a few scholastic natural philosophers abandoned the hypothetical and argued for the real existence of infinite space, thus abandoning a crucial aspect of Aristotle's cosmic picture and accepting as real what Aristotle had regarded as impossible. As mentioned earlier, those who argued in this manner were all theologians whose interest in the problem of infinite space seems to have been aroused independently of the Condemnation of

1277, although theologians were surely aware of discussions about void spaces that arose from the Condemnation.

One of the greatest of these theologians was Thomas Bradwardine (c. A.D. 1290–1349), who died while serving as archbishop of Canterbury. Bradwardine made momentous contributions to the development and acceptance of the concept of an infinite void space. In a theological treatise titled *In Defense of God Against the Pelagians,* Bradwardine used a mathematical format to organize his theological comments to present a series of steps that led him to conclude that God could have created the world in any void place he pleased. Moreover, because God could create an infinite number of void places and create the world in any one of them, Bradwardine concluded that God must be in every one of those voids, for otherwise he would have had to move to the place in which he chose to create the world. This was unacceptable because it was universally assumed that God is immutable and that motion is a sign of change and mutability. Thus, if God had to move from one place to another in order to create the world in one of an infinite number of possible places, this would imply that God is subject to change. Because this was unacceptable, Bradwardine assumed that God occupied all the possible infinite void places, which taken collectively form an infinite imaginary void space in which God is omnipresent.

How did Bradwardine relate an eternally existent infinite void space and the God who occupied it? To avoid serious difficulties, Bradwardine made God and the infinite void space he occupied one and inseparable by assuming that God is omnipresent in the infinite void space. Did his omnipresence in an infinite void space make God a three-dimensional being? It did not because, according to Bradwardine, God "is infinitely extended without extension and dimension." By denying extension to God, Bradwardine also denied extension to the infinite void space that was God's immensity. God was thus infinitely omnipresent in a dimensionless infinite void space. This was a widely accepted interpretation in the late Middle Ages. Bradwardine's infinite void space was unlike any other earlier description of a vacuum, as is apparent from his declaration that a "void can exist without body, but in no manner can it exist without God" (Grant 1974, 557; see also Grant 1996, 122–124).

Bradwardine believed that God had to be in a place to act on what was in that place. In order to act on every possible place, it was therefore necessary to assume that God is everywhere. John Duns Scotus (c. A.D. 1266–1308) and his followers disagreed. They assumed that God could act at a distance and need not be in the place he wishes to influence or affect. Therefore, it was not necessary to conclude that God must be omnipresent in an infinite void space. But Bradwardine's concept was the more influential. Nevertheless, it had one serious flaw: How could God occupy every place of an infinite void and yet lack dimensions? Was God perhaps a three-dimensional being? The answer that gained acceptance in the Middle Ages had already been formulated around 1235 by Richard Fishacre (c. A.D. 1205–1248) in his theological commentary on the *Sentences* of Peter Lombard. Fishacre argued that God remains infinitely omnipresent without being quantitatively, or dimensionally, extended, because the whole of God is entirely in every part of space, a concept that became known as the "whole-in-every-part" doctrine. Thus, God is infinitely omnipresent and yet indivisible, because he is wholly in the least possible subdivision of space. Many theologians accepted the idea that God was infinitely omnipresent in an infinite void that lay beyond our world. It was an idea that would have an impact on seventeenth-century ideas about space and God, although the space envisioned by some nonscholastic natural philosophers of the seventeenth century—especially Sir Isaac Newton—was three-dimensional, as indeed was God himself.

THE IMPACT OF RELIGION ON NATURAL PHILOSOPHY IN THE MIDDLE AGES

Up to this point, we have seen one aspect of the relations between natural philosophy and theology in the late Middle Ages, namely the impact of the Condemnation of 1277, especially Articles 34 and 49. The discussions deriving from Articles 34 and 49 were not religious in content, but they were the result of a tense relationship between natural philosophy and theology. Arts masters who became involved in questions about God's absolute powers conceded that God could do anything whatsoever, short of a logical contradiction, but the discussions were in no way of a religious character. Indeed, even the discussions about God's omnipresence in an infinite void, which were confined to theologians, were not religious or doctrinal in any way. No one had to accept Bradwardine's interpretation of God's dimensionless infinite omnipresence. It was not a doctrinal issue or an article of faith, although there were significant disagreements and counter-interpretations, as there were on so many problems in natural philosophy.

But we must now go beyond the Condemnation of 1277 and inquire about the overall influence and impact of theology and religion on treatises in natural philosophy, whether composed by arts masters or theologians. At least one scholar, Andrew Cunningham, has claimed that natural philosophy from the Middle Ages to the seventeenth century "was not just 'about God' and His creation at those moments when natural philosophers were explicitly talking or writing about God in their natural philosophical works or activities. It was, by contrast, 'about God' and His creation the whole time" (Cunningham 1991, 388). Elsewhere he declares: "Over and above any other defining feature which marks natural philosophy off from modern science . . . natural philosophy was *about God* and *about God's universe*. Indeed, this was the central pillar of its identity as a discipline; both with respect to its subject-matter and to its goals, its purposes, and the functions it served. This is what, more than anything else, distinguishes it from our modern science" (Cunningham 1991, 381). However well this may serve as a description of natural philosophy in the seventeenth century, it is false with respect to the Middle Ages.

In the course of teaching and studying Aristotle's natural books in the arts faculties of medieval universities for more than three centuries, a vast body of commentary literature was produced (this section draws heavily on Grant 1999, 243–267). The authors of those treatises, whether arts masters or theologians, firmly believed that God, who was frequently referred to as the First Cause (*prima causa*), had created the world from nothing and was also the ultimate cause of all events or effects. As Christians, of course, they also held many other religious beliefs. Did all these religious beliefs affect the way medieval scholars wrote natural philosophy? Did it mean that their objective in doing natural philosophy was essentially theological or religious, as Cunningham would have it? The response to these two questions is firmly in the negative.

To begin with, as we saw earlier, natural philosophy was a subject that belonged to the faculty of arts and was wholly independent of the faculty of theology. Given the Oath of 1272 at the University of Paris and the Condemnation of 1277, arts masters had little incentive to write about theology, unless it became unavoidable in some particular question. An examination of 310 questions on five of Aristotle's treatises (*Physics* by Albert of Saxony, 107 questions; *On*

Generation and Corruption by Albert of Saxony, 35 questions; *On the Heavens* by John Buridan, 59 questions; *On the Soul* by Nicole Oresme, 44 questions; and on *Meteorology* by Themon Judaeus, 65 questions) strongly supports this claim. In the 310 questions that appear in the five treatises that formed the core of Aristotle's natural philosophy, it becomes readily apparent that late medieval natural philosophy was basically about Aristotle's principles, opinions, ideas and concepts, and therefore about natural phenomena, not God, faith, and the supernatural. Of the 310 questions, 217 are free of any involvement with theology or faith; 93 (or approximately 29 percent) mention God and the faith. Inspection of these 217 questions would not reveal whether the author was Christian, Muslim, Jewish, agnostic, or atheist. Of the 93, with at least a trace of theological sentiment, 53 mention God, or something about the faith, in a perfunctory manner. Of the remaining 40 questions, only 10 have relatively detailed discussions about God or the faith.

When natural philosophers had occasion to invoke God in their treatises, or to mention some aspect of the faith, they usually did so in one of four categories. In the first, the natural philosophers might mention God in places where Aristotle mentioned God, the gods, or something about divinity, or where late Greek commentators reported opinions of Greek and Roman pagan philosophers on some issue that bears on Christian doctrine or faith. Such passages were relatively rare. Included in the second category are Aristotle's ideas that were contrary to the Christian faith, as when he argued that the world had no beginning and would have no end, or that accidents must inhere in a substance and cannot exist independently of substance, an idea that would have made the Eucharist, or Mass, impossible, because it required that the accidents of the bread and wine continue to exist even after the bread and wine had been miraculously transformed into the body and blood of Christ.

The third and fourth categories occurred much more frequently than the first two and are more significant. In the third category, God and articles of faith are used in an analogical sense, or by way of example, to serve as a means of comparison with natural phenomena, or simply to illustrate something about the natural world. For example, in his questions in *On the Soul*, Nicole Oresme asserts: "Some power makes this or that operation anew without changing itself, just as is obvious with God who continuously produces new effects without any change in Himself." Similarly, Themon Judaeus uses God in a comparative manner when he asserts that "a pure element is understood [to be] simple, but not simple absolutely, as is God, or an intelligence" (Grant 1999, 260).

The most striking use of God in medieval natural philosophy has already been discussed, namely his absolute power to do anything short of a logical contradiction. As we saw, the invocation of God's absolute power in natural philosophy in the late thirteenth and fourteenth centuries was a consequence of the Condemnation of 1277. Earlier, I described the impact of Articles 34 and 49. In addition to those two articles, there were numerous mentions of God's absolute power to do this or that naturally impossible act, as when John Buridan discussed the immobile empyrean heaven, regarded as the abode of the blessed and the outermost celestial sphere in the cosmos, which, according to most theologians, enclosed the world. Buridan declared of this immobile heaven that "according to nature [it] does not have any potency or inclination for motion, although it could be moved supernaturally by God Himself, just as all things, except God Himself, could be annihilated by God" (my translation from Buridan 1942, 152). Buridan's assertion had nothing to do with the Condemnation of 1277 but is merely representative of the numerous contexts within which God's absolute power was invoked. It was not unusual for medieval natural philosophers to distinguish between what God could do by his absolute power and what could be done by natural powers. When opting for the latter, they frequently used the expression "speaking naturally" (using some form of the Latin *loqui naturaliter*).

There are mentions of God in commentaries and questions on Aristotle's works that do not fall into any of the four categories I have identified. These are of minor importance: they are often little more than mere mentions of the name "God" or they assert something that God can or cannot do. These instances, and the four categories already mentioned, reveal a very modest impact on natural philosophy. The overwhelming number of questions included nothing about religion or theology. They were straightforward discussions about problems in natural philosophy.

Scholars in the Middle Ages, whether arts masters or theologians, had no desire to Christianize Aristotelian natural philosophy. The views of Albertus Magnus and his student, Thomas Aquinas, described in chapter 6, apply equally to scholars in the fourteenth and fifteenth centuries. Like Albertus, they all sought to avoid the intrusion of theology and religious matters into natural philosophy. No one exemplified this approach better than John Buridan, who sought to explain nature's operations in terms of natural causes and effects, and not to explicate God's supernatural actions and miracles. Buridan had no problems with his faith. He accepted the truths of revelation as absolute. But when doing natural philosophy, he simply assumed that his task was to explicate problems about natural phenomena and not to deal with the supernatural. In *Questions on On the Heavens* (Book 1, question 25), Buridan asks whether every generable thing will actually be generated, then responds that this question can be treated naturally—"as if the opinion of Aristotle were true concerning the eternity of the world, and that something cannot be made from nothing" (Grant 2001, 198)—or supernaturally, with the explanation that God could prevent a generable thing from generating naturally by simply annihilating it. "But now," Buridan declares, "with Aristotle, we speak in a natural mode, with miracles excluded" (Grant 2001, 198). If he had to concede that God could use his absolute, unpredictable power to produce any natural impossibilities he wished, Buridan could still save Aristotle and natural philosophy by characterizing Aristotle's arguments as sufficient in the real, natural world, the one he and his fellow natural philosophers sought to understand and explain.

THE ROLE OF NATURAL PHILOSOPHY IN THEOLOGY

If theology and faith exerted only a small, superficial influence in natural philosophy and played no significant role in shaping that discipline, what role did natural philosophy play in the development of theology as it was studied and written about in schools of theology, most importantly in the school of theology at the University of Paris?

In chapter 5, I briefly described the genre of theological literature known as *Commentaries on the Sentences of Peter Lombard.* Each student in the school of theology was expected to comment on the books of Peter Lombard. Although Peter's books were not in the form of questions, the commentaries on his work were indeed subdivided into questions, following the same format as questions treatises on Aristotle's natural books. The commentaries were originally given as lectures and then frequently "published," in the sense that written versions were produced by each author, and copies were eventually made and disseminated, many of which have been preserved, and many of which have perished but are known to have once existed. The lectures and their written counterparts were often divided into distinctions, articles, and questions. The questions are easy to detect because, like most of their counterparts in natural philosophy, they begin with the word "whether" (*utrum*). Over the course of a four-book commentary, a great number of questions were posed. For example, in the thirteenth century, Richard of Middleton presented 1,862 distinct questions and Thomas Aquinas, approximately 1,700. Other authors considered fewer questions, in the neighborhood of 500 to 600. It is hardly surprising, therefore, that the average *Commentary on the Sentences* was quite lengthy.

In the course of the fourteenth century, the most striking feature of the *Sentence* commentaries is that their authors reveal less and less concern for religion and more and more of an obsession with natural philosophy and logic. They were captivated by a largely analytical approach to theological problems. Indeed many theological questions became little more than vehicles for the practice of natural philosophy, or for coping with problems in logic. Theology became an analytic discipline. Church authorities tried to stem the tide, but their efforts were a dismal failure. Popes Gregory IX (A.D. 1227–1241) and Clement VI (A.D. 1342–1352) criticized theologians for being overly absorbed in philosophical questions. In 1366, the University of Paris attempted to separate theology and natural philosophy as much as was feasible. All these efforts failed because, in the final analysis, the church authorities, most of whom were themselves trained in natural philosophy and logic, were fully aware that it was necessary to use those disciplines to study theology. The futility of these efforts is made apparent by John Major, a sixteenth-century theologian, who explained that "for some two centuries now, theologians have not feared to work into their writings questions which are purely physical, metaphysical, and sometimes purely mathematical" (Grant 2001, 281–282).

The influence of natural philosophy on theology is made most apparent by the numerous questions routinely discussed in treatises on Aristotle's natural philosophy, which were also frequently included in *Sentence* commentaries, especially in the second book, which was devoted to the biblical creation. Questions of the following kind were fairly common:

whether the heaven is composed of matter and form;

on the number of spheres, whether there are eight or nine, or more or less;

whether the heaven is spherical in shape;

whether the heavens are animated;

whether the whole heaven from the convexity of the supreme [or outermost] sphere to the concavity of the lunar orb is continuous or whether the orbs are distinct from each other;

whether celestial motion is natural;

whether the stars are self-moved or are moved only by the motions of their orbs. (Grant 2001, 275–276)

Many other similar questions appeared in both types of treatises. There were also purely secular questions that appeared only in theological commentaries. An even more widespread tendency in theology was the application of analytic techniques—drawn from logic, natural philosophy, and mathematics—to various theological questions. Analyses using these instruments of reason were apt to turn up in any theologian's commentaries on any one of the four books of the *Sentences*. But the most interesting questions occurred in the first and second books, the former devoted to God, and the latter to creation and angels, among other topics.

Questions about God in the first book usually inquired about his power, his knowledge, what he could or could not do, and what he had intended to do. For example, commentators raised such questions as:

whether God could have made the world before He made it;

whether God knew that He would create a world from eternity;

whether God could do evil things;

whether the Creator could have created things better than He did;

whether God could make a better world than this world;

whether God could make the future not to be;

whether God could make a creature exist for only an instant. (Grant 2001, 359, 251)

There were also questions that sought to discover the effects on God if he changed his mind, as in these two questions:

whether, without any change in Himself, God could not want something that at some [earlier] time He had wanted; and

whether if God wants something new that He did not want from eternity, would this constitute a real change [in God]. (Grant 2001, 359)

God was always assumed capable of doing anything that did not involve a logical contradiction. In the thirteenth century, Richard of Middleton asked "whether God could do contradictory things simultaneously" and formulated this almost universally held judgment. "I respond," he declares,

that God cannot make two contradictories exist simultaneously, not because of any deficiency in His power, but because it does not make any sense to [His] power in any way. And if you should ask why this does not make possible sense, it must be said that with respect to this [problem] no other argument can be given except that such is the nature, or the disposition, of affirmation and negation, just as if we sought why every whole comprehends a part, no other argument would be forthcoming than that such is the nature of whole and part. (Grant 2001, 227)

In another question, in the same book and distinction, Richard asks "whether God could be called omnipotent because He can make all things that He wishes to make." In response, Richard declares:

God is omnipotent and can make all things that He wishes. This power to make all things that He wishes to make is not the precise reason of His omnipotence. But He ought to be called omnipotent because He is able to make everything that is absolutely possible, that is possible as was said in the preceding question, and this applies to everything that does not include a contradiction. (Grant 2001, 227–228)

The way in which the law of non-contradiction determined whether God could do this or that action is nicely illustrated by a question drawn from the *Sentence Commentary* of Gregory of Rimini (d. A.D. 1358). Among the questions Gregory asks about God is whether God could make someone, say Peter, sin. "If God would make Peter sin," Gregory argues, "Peter would sin and not sin. That he would sin [if God wanted him to sin] is obvious. But that he would not sin is now proved: because no one sins by doing what God wishes him to do or makes him do. . . . It is therefore obvious, properly speaking, that God cannot make Peter sin" (Grant 2001, 225).

Medieval natural philosophers became engrossed in problems about infinites—the infinitely large and the infinitely small. Much of their interest was focused on the infinite divisibility of a mathematical continuum. Theologians, who frequently used mathematical concepts developed within natural philosophy, linked this interest to various possible actions by God. They made use of their knowledge that there are an infinite number of proportional parts in an hour. Gregory of Rimini used this knowledge to describe how God could make an actually infinite multitude of angels in an hour. In the first proportional part of the hour, God creates an angel and preserves it. He creates another angel in the second proportional part of the hour and preserves it, and then creates and preserves an angel in every successive proportional part of the hour. Because there are an infinite number of proportional parts in an hour, God would have created an actually infinite multitude of angels at the end of the hour.

Similarly, Robert Holkot, a fourteenth-century English theologian, combined the infinite divisibility of the mathematical continuum into proportional parts with common interest in another doctrine known as "first and last instants." Holkot applied all this to an example involving free will, merit, and sin. He imagined a situation in which a man is alternately meritorious and sinful during the last hour of his life. The man is thus meritorious during the first proportional part of his final hour and sinful in the second proportional part; he is again meritorious in the third proportional part of his last hour and sinful in the fourth proportional part; and so on, through the infinite series of decreasing proportional parts until the last instant is reached and death occurs. Because the instant of death cannot, and does not, form a part of the infinite sequence of decreasing proportional parts of the last hour of the man's life, it follows that there cannot be a last instant of his life, and therefore no final instant in which he is either meritorious or sinful. Under these circumstances, God cannot know whether to reward or punish this man in the afterlife. In this case, then, the doctrine of free will reaches a limit and not even God can overcome the dilemma inherent in this example. We see that an appropriate outcome is not possible in every case involving free will and a divine system of rewards and punishments (Grant 1996, 154).

We saw earlier that medieval theologians often assumed the existence of an infinite void space beyond our world in which God is omnipresent. Their interest in the infinitely large extended far beyond spatial considerations. They repeatedly asked whether God could make something that is actually infinite, distinguishing at least three kinds of infinites: (1) an infinite magnitude; (2) an infinite multitude of things; and (3) an infinitely intense quality like hot or cold. With respect to an actual infinite to which nothing more could be added because there was nothing left to add, theologians were divided in their response as to whether God could create such an infinite. Many theologians thought God could create an actual infinite, but they could not explain a paradoxical element that received no answer during the Middle Ages. The non-theologian John Buridan expresses the paradox in his *Questions on Aristotle's Physics* (Book 3, question 15) when he asks "whether there is some infinite magnitude?" Buridan denies that God could create an actually infinite magnitude, because if he did, "he could not create anything that is greater, since it is repugnant [or absurd] that there should be something greater than an actual infinite" (Grant 1994, 111). Thus, if there is nothing bigger than an actual infinite, it is a contradiction to suppose that God could create something greater than an actual infinite. But if he cannot create anything bigger than an actual infinite, God's absolute power would be restricted. Buridan was merely following the general rule that God can do anything short of a logical contradiction. Because a contradiction is involved in this situation, Buridan merely followed the path that any theologian would have taken. Nevertheless, he was wary about treading into the theological domain, and he concludes his argument by declaring that "with regard to all of the things that I say in this question, I yield the determination of them to the lord theologians, and I wish to acquiesce in their determination" (Grant 1994, 112).

Theologians also found occasions to argue about whether one infinite could be greater than another infinite, or if instead all infinites are equal. Saint Bonaventure raised some of the problems associated with infinites in his attack on the Aristotelian assumption of an eternal world. In the first of six arguments against an eternal world, Bonaventure believes he detects a contradiction when he compares the revolutions of sun and moon in a world that has an infinite past. Drawing upon the arguments of John Philoponus (see chapter 4), Bonaventure asserts that the moon will circle the earth twelve times a year, and the sun only once. It follows, Bonaventure argues, that over an infinite past, the moon will have made twelve times as many revolutions as the sun, from which it further follows that the infinite number of revolutions made by the moon will be greater than the infinite number of revolutions made by the sun. Because Bonaventure, like many others, assumed that all infinites are equal, he concluded that if an eternity of time could produce such a contradictory consequence, it must be false.

In the fourteenth century, attitudes toward the actual infinite changed considerably. Bonaventure's assumption that one infinite can exceed another was rejected by other theologians. If the world existed from all eternity, says Robert Holkot, it followed that some—and this would have included Saint Bonaventure—believed that "there would be a greater number of fingers than men, and a greater number of revolutions of the moon than of the sun." Holkot rejects this reasoning and says that one must simply deny these claims: "With a thousand men, there is a greater number of fingers than men; but with an infinity of men, there is no greater number of fingers than men, because there is an infinity of men and an infinity of fingers" (Grant 2001, 243). Thus, Holkot denied Bonaventure's argument that the eternity of the world is an absurdity because a consequence of eternity would be the generation of unequal infinites. Holkot argued that the infinites deriving from an eternal world would be equal, not unequal as Bonaventure claimed. Consequently, if the world did indeed exist from eternity, the unequal infinites that Bonaventure detected would not materialize and there would be no contradiction.

Indeed, Gregory of Rimini went beyond all of his colleagues and proposed a momentous idea on the relationship between infinites, an idea that lies at the heart of the modern-day theory of infinite sets. In the course of discussing infinites, Gregory had occasion to discuss terms such as "part," "whole," "greater than," and "less than." Gregory arrived at the dazzling idea that in the domain of the infinite, a part can equal the whole. An example that is commonly used to illustrate what Gregory intended is the relationship between the natural numbers and its subset of even numbers. Because the even numbers can be placed in a one-to-one correspondence with the set of natural numbers, it follows that there are as many even numbers as natural numbers. Therefore, despite the fact that the set of even numbers is a part of the set of natural numbers, the two infinite sets are equal, because they can be placed in a one-to-one correspondence. Gregory had discovered the counterintuitive idea that in the domain of the infinite, a part can equal the whole. Unfortunately, Gregory's discovery had no further consequences in the Middle Ages. As John Murdoch explains,

Gregory's resolution of the paradox so frequently generated by the assumption of the eternity of the world is by far the most successful and impressive I have discovered in the Middle Ages. Unfortunately, however, it appears hardly ever to have received the understanding and appreciation it clearly deserved. Since the "equality" of an infinite whole with one or more of its parts is one of the most challenging, and as we now realize, most crucial aspects of the infinite, the failure to absorb and refine Gregory's contentions stopped other medieval thinkers short of the hitherto unprecedented comprehension of the mathematics of infinity which easily could have been theirs. (Murdoch 1969, 224; also cited in Grant 2001, 248)

Murdoch further explains that although medieval scholars failed to capitalize on Gregory's extraordinary foresight and insight, such problems about the infinite were common to medieval natural philosophy and theology and "led them, as it seldom did their ancient predecessors, to the heart of the mathematics of the infinite. The fact that they seemed to realize that it was the heart, and that in treating it they fared as well as, and at times better than, anyone else before, it appears, the nineteenth century, is unquestionably to their credit" (Murdoch 1969, 224; also cited in Grant 2001, 248).

The inclusion of numerous questions devoted to the existence and attributes of angels in theological commentaries is hardly surprising. Indeed, it is what we would expect to find in theological treatises, but not in questions on Aristotle's natural philosophy, in which angels were rarely ever mentioned and never discussed. Most questions about angels concerned their moral behavior. For example, Thomas Aquinas, in his commentary on the *Sentences*, asked "whether created angels are blessed," "whether angels were created in grace," "whether in angels there could be sin" (Grant 2001, 255), and so on. These were typical questions asked in commentaries on the *Sentences* of Peter Lombard. But apart from their spiritual importance, angels were made to serve another role: they became a convenient vehicle for the importation of natural philosophy into theology. This angelic influx of natural philosophy into theology took the form of questions about the motion of angels: the way they occupied a place and the way they moved from place to place (see Figure 7.2).

In *Summa of Theology*, Thomas Aquinas considered the relationship of angels to places and, in the form of articles, asked the following three questions: "does an angel exist in a place?"; "can an angel be in several places at once?"; and "can several angels be in the same place at once?" Thomas inquired next about "the local motion of angels," and under this theme posed three more questions: "Can an angel move from place to place?"; "Does an angel, moving locally, pass through an intermediate place?"; and finally, "whether an angel's motion occurs in time or in an instant?" (Grant 2001, 256). With such questions, theologian–natural philosophers were led to compare the motions of material bodies, as Aristotle described them, with the motions of the immaterial substances of angels, and to determine whether an angel occupied a place in the same way as did a material body, or in some other manner. Generally speaking, angels were assumed to occupy places quite differently than did material bodies. Bodies were co-extensive with the places they occupied, whereas an-

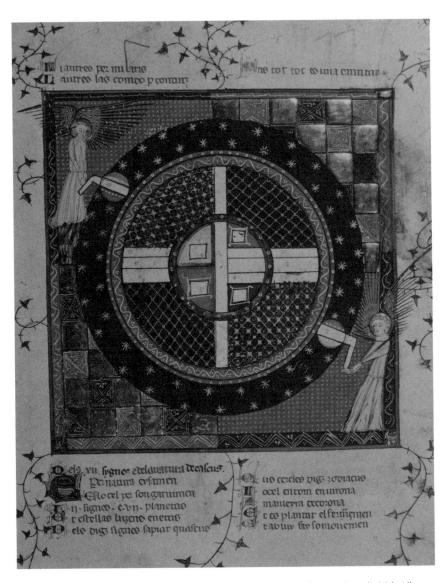

Figure 7.2. Angels cranking the world to produce its circular motion. (British Library, Harl. MS. 4940, fol. 28. See Murdoch 1984, 336, for a different illustration of angels cranking the world.)

gels were not, but merely delimited by the places they occupied. Moreover, because an angel was regarded as indivisible, theologians assumed that the whole angel could be in every part of the place that delimits it. As for the motion of angels from one place to another, the questions posed by Thomas Aquinas, and cited above, were commonly discussed—namely, whether an angel's motion is instantaneous, and whether an angel passes through the mid-point of its path before it reaches its destination. Most rejected instantaneous motion. Richard of Middleton, for example, denied that an angel could move through any medium from one place to another in an instant. For if it could, God, the most powerful force in existence, should be able to move that angel the same distance in less than an instant. But, argued Richard, not even God could move an angel through any distance in less time than an instant, because there is no temporal measure smaller than an instant.

Interest in first and last instants of a process was widespread among scholastic natural philosophers. Theologians eagerly imported such considerations into their own questions. We saw earlier that Robert Holkot sought to determine if God could know whether a man died in a meritorious or sinful state in the last instant of his life. Similarly, Hugolin of Orvieto asked whether angels, in the first instant of their existence, could be freely unmeritorious and also meritorious (Grant 2001, 259).

Not infrequently, theologians transformed what purported to be a discussion of some problem about angels into a lengthy consideration of time, the mathematical continuum, indivisibles, problems of motion, the infinite, and generally, problems that involved the application of logico-mathematical techniques. One of the most extraordinary ostensible discussions about angels appears in Book 2 of Gregory of Rimini's *Commentary on the Sentences.* In a question of approximately 42 pages in length in the modern edition, Gregory asks "whether angels were created before time [began], or after time [began]" (Grant 2001, 259). He mentions angels at the beginning of the question but does not mention them again for approximately 39 pages. Instead of discussing angels, Gregory presents a detailed analysis of time, drawing heavily on the philosophical works of Aristotle and Averroes. In the very next question, Gregory asks "whether an angel exists in a divisible or indivisible place" (Grant 2001, 260). Gregory divides this question into two articles, in the first of which, some 53 pages in length, he mentions angels only once. Only in the second article, which is a mere 9 pages in length, does Gregory concern himself with angels.

In the first lengthy article, Gregory deals with three separate conclusions that involve mathematics, physics, and logic. In the course of this extensive section, Gregory cites Euclid's *Elements* numerous times and includes fourteen geometrical diagrams. He includes additional questions on angels, much of which are devoted to problems of motion, the kind of problems that were normally considered in treatises on natural philosophy. In the fourteenth century, many theologians followed Gregory of Rimini's approach. They converted theological problems into exercises on natural philosophy, mathematics, and logic.

Natural philosophy was even applied to the Eucharist and Mass, central articles of faith in the Catholic Church. In the Eucharist, God miraculously produces the actual body and blood of Christ in the bread and wine used in the Mass. Although the bread has become the substance of the body of Christ, the bread retains its usual visible accidental properties. Despite appearances, however, the accidental properties no longer inhere in the substance of the bread; they only behave and appear as if they do. Nor do they exist in Christ. These apparently free-floating accidents violated a fundamental doctrine of Aristotle's physics, namely that all accidents, without exception, must inhere in a substance. Because it conflicted with the Eucharist, three successive articles in the Condemnation of 1277 denounced the idea that an accident cannot exist unless it inheres in a substance. The condemned articles declared:

139. That an accident existing without a subject is not an accident, except equivocally; [and] that it is impossible that a quantity or dimension exist by itself because that would make it a substance.

140. That to make an accident exist without a subject is an impossible argument that implies a contradiction.

141. That God cannot make an accident exist without a subject, nor make several dimensions exist simultaneously [in the same place]. (Grant 1996, 78)

It was the task of medieval theologians to explain how the accidents could appear as if they inhered in the bread but did not, and how this was to be reconciled, if at all, with Aristotle's doctrine. Theologians gave a range of responses. Some, like William of Ockham tried to stay as close to Aristotle's explanations as was possible or feasible. As Edith Sylla has explained, Ockham chose not to modify Aristotle's physics in order to explain what seemed inexplicable by natural philosophy, but explained the naturally inexplicable by assuming God's direct intervention (Sylla 1975, 363). Thus, Ockham chose to invoke God's direct supernatural intervention at those points where he would have had to twist natural philosophy to fit theological requirements. As Sylla expresses it, Ockham "allows natural philosophy its proper autonomy" (Sylla 1975, 366).

In purely theological questions, such as those about the Eucharist, many scholars, like Ockham, did not use natural philosophy to explain the manifestly supernatural actions, which they regarded as beyond reason. But they used natural philosophy to explain various aspects of a theological problem that were amenable to rational discourse. In questions about the Eucharist, for example, theologians discussed qualities, quantities, and the nature of dimensions, topics that Aristotle had discussed and which his followers elaborated upon. Article 139, cited above, condemned the Aristotelian axiom that no quality, quantity, or dimension could exist independently of a material body, and Article 141 denounced a basic Aristotelian concept that two or more dimensions could not exist simultaneously in the same place. These Aristotelian ideas were contrary to two basic concepts inherent in the Eucharist, and both had to be rejected if the Eucharist was to be made credible to the faithful. Not only did Articles 139 and 141 place limitations on God's absolute power, but Article 139 denied that God could miraculously produce the actual body and blood of Christ in the bread and wine used in the Mass (several different dimensions could not exist in the same place simultaneously). Articles 139 and 141 both denied that the accidents of the bread, though still visible, could inhere in the substance of the bread, which had become the body of Christ.

For better or worse, natural philosophy became embedded in such theological themes as the Eucharist and the Resurrection, and in such problems as the increase of charity, or grace, which gave rise to a large body of literature called the "intension and remission of forms or qualities," in which the intensions and remissions of all sorts of qualities were mathematized, both arithmetically and geometrically.

Anything that could be identified as a variable quality would have been eligible for mathematization. We need only mention sensible qualities such as color and abstract qualities such as justice, health, honor, joy, and pain to convey an idea of the range of qualities. The intension of any quality involved a greater intensification of that quality, as when an apple becomes redder; or a leaf becomes greener; or a pain becomes greater. A quality undergoes remission when it becomes less intense, as when, to use the previous examples, the apple becomes less red; the leaf becomes less green; and the pain is gradually or quickly reduced. Scholastics came to assume that qualities were like extended magnitudes or weights, so that identical qualitative parts could be added or subtracted. Indeed, they arbitrarily assigned degrees to qualities and added or subtracted them, thereby making qualitative variation amenable to mathematical treatment, using both arithmetic and geometry. The medieval origin of the problem is found in Peter Lombard's *Sentences,* Book 1, distinction 17, in which Peter asks "whether it ought to be conceded that the Holy Spirit could be increased in man, [that is] whether more or less [of it] could be had or given" (Grant 1996, 99). Peter replies that charity, which is caused by the Holy Spirit, cannot increase in man, because that would imply a change in the Holy Spirit, which is unchangeable. Hence, if charity varies from person to person, it is because humans are capable of a greater or lesser participation in the unchanging Holy Spirit.

Because from the thirteenth century onward all theological students commented on the *Sentences,* Peter's discussion of variations in grace served to stimulate interest in the problem of qualitative variation— or the intension and remission of forms, or qualities, as they came to be called. It is from these beginnings that an important body of medieval literature came into being on the variation of qualities. The intension and remission of forms became the most mathematized aspect of medieval natural philosophy. Although the problem continued to be discussed in the *Sentence Commentaries,* the most important work was done in treatises on natural philosophy. The quantitative and mathematical description of qualitative variation was first developed at Merton College at Oxford University during the 1330s and 1340s, and soon reached the University of Paris, where sometime around 1350, Nicole Oresme composed the most comprehensive and extraordinary treatise on qualitative variation (see Figure 7.3). Because velocities were treated as analogous to qualities, Oresme included, among a large number of mathematical proofs, a geometric proof of what was later called the "mean speed theorem"—that is, a proof of the mathematical formulation $s = 1/2at^2$, where s is distance traversed, a is acceleration, and t is time. Although Oresme did not recognize the importance of this law of falling bodies, Galileo did when he applied

Figure 7.3. Nicole Oresme in his study. The instrument is an armillary sphere used in the study of astronomy. Miniature from Oresme's French translation of Aristotle's Latin text of *On the Heavens*. (Bibliothèque Nationale, MS. Fr. 565, fol. 1r.)

it to falling bodies in the sixteenth century. Thus did theology provide a springboard for the development of an important medieval subdiscipline that involved the extensive application of mathematics to natural philosophy.

THE SIGNIFICANCE AND MEANING
OF THE INTERACTION BETWEEN NATURAL
PHILOSOPHY AND THEOLOGY

From all that has been said about the interrelations between natural philosophy and theology, it is obvious that theology had a relatively small impact on natural philosophy, whereas natural philosophy, logic, and mathematics had so great an influence on theology that they reshaped the discipline, transforming its subject matter more nearly into natural philosophy than theology or religion. "One can point to numerous *Sentence Commentaries*," observes Edith Sylla,

in which natural science is used extensively, and there are some *Sentence Commentaries* which in fact seem to be works on logic and natural science in disguise—in response to each theological question raised, the author immediately launches into a logical-mathematical-physical disquisition and then returns only briefly at the end to the theological question at hand. (Sylla 1975, 352)

To Sylla's pertinent comment, we may appropriately add one from John Murdoch, who declares that "genuine parts of fourteenth-century theological tracts . . . successfully masqueraded as straightforward tracts in natural philosophy" (Murdoch 1975, 276).

It appears that natural philosophy in the hands of theologians became more innovative and exciting than the natural philosophy produced by arts masters, who were non-theologians. Although arts masters, like John Buridan, pursued questions involving God's absolute power, they were more apt than theologians to emphasize that although God could do all those acts, it did not follow that he had done them. Although theologians would probably have agreed with the arts masters, they treated most of the counterfactual questions as real problems to be resolved by the use of all available tools, which almost always meant natural philosophy, logic, and mathematics. And because they were applying natural philosophy to interesting, and even bizarre, theological questions, the theologians were extending natural philosophy into many areas that the arts masters were, for all practical purposes, forbidden to enter.

Thus, the potential intermingling of theology and religion with natural philosophy was the business of theologians, not natural philosophers, who lacked theological training. The consequences of this approach for theology were significant. Many of the questions theologians included in their *Sentence Commentaries* give the appearance of having been formulated so that the authors might apply or test their reasoning skills by means of the basic analytic tools available to them—namely, natural philosophy, logic, and mathematics. Otherwise, why would they have asked questions such as:

whether an angel is in a divisible or indivisible place;

whether an angel could sin or be meritorious in the first instant of his existence;

whether God could cause a past thing [or event] to have never occurred;

whether an angel could be moved from place to place without passing through the middle [point]. (Grant 2001, 277–278)

All of these, and many more like them, required analytic techniques to produce plausible responses. Medieval theologians were apparently strongly motivated—or perhaps even compulsively driven—to explain as much about their faith in rational terms as was feasible. As theologians made theology more and more analytic, they seem to have emptied their discipline of spiritual content. All attempts by the church to make theology less analytic and more spiritual failed dismally. Indeed, the attempts were often half-hearted.

The hundreds of questions on natural philosophy in treatises on Aristotle's natural books and in works on theology probed all aspects of the world, including physical nature, the supernatural, and an imaginary world of hypothetical and possible phenomena. The only significant restrictions on inquiry into natural phenomena during the late Middle Ages in both natural philosophy and theology were related to revelation. Revelation was comprised of truths derived directly from God, or from his revealed word in holy scripture. Revelation embraced such fundamental Christian truths as the Trinity, incarnation, redemption, the Eucharist, and the divine creation of the world from nothing. These were regarded as beyond the comprehension of reason and logic and were to be believed on the basis of faith alone. Most theologians who attempted to analyze revealed truths usually did so to better comprehend, or explain, what they already believed on faith. If, perchance, a scholar arrived at conclusions that were contrary to revealed truth, he had to reject them or be accused of heresy. One could, however, discuss propositions or conclusions that were the opposite of revelatory truths. That is, one could simply assume their truth hypothetically for the sake of an argument. Indeed, that is how Aristotle's contrary-to-faith doctrines were discussed, especially after the Condemnation of 1277. For example, one could set aside the doctrine of creation from nothing by assuming in natural philosophy that material bodies could come into being only from previously existent material bodies. While such an assumption was made by all medieval Christian natural philosophers, they would all have agreed that while it was true for natural things, it was false in the Christian religion, in which it was an article of faith that the world was created from nothing.

But what about holy scripture? How were its texts to be interpreted by natural philosophers, especially when these might conflict with assertions regarded as true in Aristotle's natural philosophy? On this important theme, general guidelines were furnished by Saint Augustine in his commentary on Genesis: "In matters that are obscure and far beyond our vision, even in such as we may find treated in Holy Scripture, different interpretations are sometimes possible without prejudice to the faith we have received. In such a case, we should not rush headlong and so firmly take our stand on one side that, if further progress in the search of truth justly undermines this position, we too fall with it" (Augustine 1982, 1:41). Augustine did, however, insist that whenever possible and feasible, holy scripture should be taken literally, as, for example, in interpreting the waters above the firmament. "Whatever the nature of that water and whatever the manner of its being there," he declares, "we must not doubt that it does exist in that place. The authority of Scripture in this matter is greater than all human ingenuity" (Augustine 1982, 1:52).

In *Summa of Theology*, Thomas Aquinas adopted both aspects of Augustine's approach, as, indeed, did Saint Bonaventure. In discussing whether the firmament was made on the second day, Thomas declares (pt. 1, question 68, art. 1):

Augustine teaches that two points must be kept in mind when resolving such questions. First, the truth of Scripture must be held inviolable. Secondly, when there are different ways of explaining a Scriptural text, no particular explanation should be held so rigidly that, if convincing arguments show it to be false, anyone dare to insist that it is still the definitive sense of the text. Otherwise, unbelievers will scorn Sacred Scripture, and the way to faith will be closed to them. (Thomas Aquinas 1967, 10:71–73)

Thomas' attitude proved popular and shaped the attitude of numerous theologians toward the Bible and its relations to natural philosophy. Virtually no theologians or natural philosophers from the late thirteenth to fifteenth centuries interpreted biblical descriptions of cosmological or physical phenomena in a literal sense. All scholastic natural philosophers and theologians accepted Aristotle's judgment that the world is a sphere, despite biblical statements in Psalm 103 in which heaven is said to be stretched like a skin, or to be like an arched roof. In his commentary on the *Sentences*, Saint Bonaventure explains the conflict between the natural philosopher's interpretation of the shape of the world and the biblical descriptions by informing his readers that "Scripture, condescending to poor, simple people, frequently speaks in a common way. And so when it speaks about the heaven, it speaks in a way that the heaven appears to our senses, and [therefore] says that with respect to our hemisphere the heaven is like a skin, or a stretched, arched roof" (slightly altered from Grant 2001, 266).

In the course of presenting arguments about the possible daily axial rotation of the earth, Nicole Oresme, in the fourteenth century, adopted a similar attitude. To silence those who cited biblical passages clearly indicating that the sun moved around an immobile earth, Oresme observed that one could reply "by saying that this passage conforms to the customary usage of popular speech just as it does in many other places, for instance, in those where it is written that God repented, and He became angry and became pacified, and other such expressions which are not to be taken literally" (Oresme 1968, 531).

By adopting a flexible attitude about the interpretation of biblical passages, the Church and its theologians inadvertently made a highly significant contribution to the advancement of science and natural philosophy. Had they insisted on a literal interpretation of scripture, rather than allow for allegorical and metaphorical interpretations, natural philosophy would probably have been stifled and retarded. Over and over again, literal interpretations were replaced by interpretations from natural philosophy for which reasonable evidence could be furnished. Despite the debacle in the seventeenth century when the

church persecuted Galileo for adopting Nicholas Copernicus' heliocentric, or sun-centered, system instead of Aristotle's geocentric
world, Christians were largely convinced that the Bible was not a book
about science. In Islam, by contrast, the Qur'an was, and is, interpreted literally, because it is regarded as the direct word of God, dictated to Muhammad by the angel Gabriel. Some Islamic scholars are
convinced that there is much of scientific value in the Qur'an and that
it "contains many observations on natural phenomena and explana-
tory details in total agreement with modern scientific discoveries"
(Shakir 1999, preface). Despite occasional fantastic modern claims for
the Qur'an's scientific significance (Hoodbhoy 1991, 140–154), it is no
more relevant to science than are the Christian and Jewish scriptures.

The realization that it would be unwise to convert the Bible into a
book that allegedly contained the secrets of nature and its operations
served to free natural philosophy and science to pursue their own
paths of investigation, unencumbered for the most part by the need
to interpret biblical texts.

What is rather amazing in the Middle Ages is the fact that theologians, or more appropriately theologian–natural philosophers, as well
as natural philosophers among the arts masters at the universities, employed a questioning approach toward a wide range of problems involving nature, the supernatural, and hypothetical and imaginary
conditions. They were as interested in the way things might have been
if God had made the world differently, or if he had created other
worlds, as they were about phenomena in the real world. This inquiring spirit was a major development in the Western world during
the Middle Ages. It was, as I have described it elsewhere, a spirit of
"probing and poking around" (Grant 2001, 361–363). What is perhaps
most astonishing about this new spirit is that it was dominated by theologians who employed the analytic skills they acquired at the universities to cope with questions and problems in natural philosophy
as well as in theology. Indeed, they made theology a large repository
of analytic techniques comprised of natural philosophy, logic, and
mathematics. All this occurred within the culture of Western Europe
after Europe emerged from centuries of decline during the barbarian
invasions. But why did this occur in Western Europe and not in the
Islamic world, or in the Byzantine Empire, home of the Greek Orthodox Church? In the concluding chapter, I shall attempt to provide a
plausible answer.

Chapter 8

⊷ ≍◊≍ ⊷

Relations between Science and Religion in the Byzantine Empire, the World of Islam, and the Latin West

We have now described in detail the long history of the relations between science and religion as it developed in Latin Christendom, or, more generally, Western Europe. Developments in Western Europe during the late Middle Ages, from approximately A.D. 1150 to 1500, were the end process of a lengthy evolution derived from two other great and important civilizations: the Byzantine Empire and Islam. It will be illuminating, and even essential, to compare the fate of science–religion relationships in these two civilizations with what we have described thus far for the medieval West. It is from these three disparate contemporary civilizations, each distinguished by use of a different language—Greek in the Byzantine Empire; Arabic in the dominion of Islam; and Latin in Western Europe—that we derive the firm foundations of Western civilization after the Middle Ages, foundations that have, for better or worse, brought us to the present. Because the Byzantine Empire was the direct successor of the Roman Empire and survived, albeit in much diminished circumstances, until it fell to the Ottoman Turks in 1453, it is appropriate to begin our comparisons with the Byzantine Empire.

THE BYZANTINE EMPIRE

From the foundation of the Roman Empire in 27 B.C.—when Octavius assumed the title of Augustus and was thereafter known as Caesar Augustus, the first Roman emperor—until the fourth century

A.D., Rome was the capital of a vast geographical area that stretched from the coastal regions along the Atlantic Ocean to the lands bordering the eastern Mediterranean. After relative peace for the first two centuries of its existence, the Roman Empire fell into turmoil and internal conflict during the next three centuries. Eventually the emperor Diocletian (A.D. 284–305) divided the empire in two, a western and eastern part, the former embracing the Latin-speaking regions of the empire, the latter the part in which Greek was the dominant language. Thus it was that in A.D. 286, the emperor Diocletian appointed Maximian, one of his trusted generals, to rule as co-emperor in the west, while he ruled in the east. Although this split did not work very well, it remained in effect until the late fifth century and then in name only until A.D. 800, when, on Christmas day, Pope Leo III crowned Charlemagne Holy Roman Emperor.

The split Diocletian made between East and West was strongly reinforced by Constantine the Great, Roman emperor from 313 to 337. In the year 330, Constantine moved his capital from Rome to the city of Byzantium, on the Bosphorus, the narrow body of water that separates Europe from Asia. The new capital was named after the emperor and called Constantinople, which, after the Turks conquered it in 1453, became Istanbul, the capital of modern Turkey. As Rome gradually emerged as the headquarters of the Roman Catholic Church, Constantinople became the great center of the Roman Empire. Although in theory, the Roman Empire was one empire under two separate but equal rulers, the western part fell into a chaotic state, and its fate was virtually independent of the eastern empire, which continued on with what was left of the Roman Empire until its fall in 1453.

Constantine the Great also played a significant role in the triumph of Christianity when, in 313, he issued the Edict of Milan (or the Edict of Toleration), granting Christians full equality with other religions, and thus freedom of worship. In 392, the emperor Theodosius made Christianity the state religion of the empire, its only legal religion. Pagan temples were closed and it became illegal to worship pagan gods. To do so was regarded as treason. Thus it was that as the empire in the West weakened and gradually declined, the church grew in strength and numbers and became the dominant force in Western Europe.

Following the split in the Roman Empire itself, there followed a split within Christianity, producing a Greek Orthodox Church in the eastern half of the empire and a Roman Catholic Church in the western part. This dramatic event occurred in the sixth century, when the Catholic Church changed the Nicene Creed that had been proclaimed in A.D. 325. Where the creed had originally declared that the Holy Spirit proceeded "from the Father" alone, the Catholic Church added the words "and the Son" (*filioque*). The Greek Orthodox Church found this highly objectionable, regarding it as tantamount to a declaration that the Holy Spirit proceeded from two distinct Gods. Thus was the split between the western and eastern Christian churches begun and carried on for centuries until it became a de facto reality in 1054, when the Pope sent legates to Constantinople who excommunicated the patriarch of the Greek Orthodox Church, along with his associates. The officials of the Orthodox Church responded similarly, by condemning the papal legates.

In the relations between church and state, the Byzantine Empire and the West differed in a major way. Where the West, as we saw, recognized a difference between church and state, the Byzantine Empire was essentially a theocracy. The Byzantine Emperor was regarded as the head of church and state. He had the authority to appoint and remove the patriarchs of the Orthodox Church. Despite their great authority, emperors rarely tried to change church dogma, failing on the few occasions when they tried. Although this lack of separation of church and state was a hindrance to the development of science and natural philosophy, the Byzantines confronted an even greater problem than dwelling in a theocratic state. The Byzantine Empire was always surrounded by actively hostile enemies and was constantly at war. So ongoing and unrelenting were the wars, that "neither literature nor science benefited from those strong impulses which they normally derive from the human energies that are released in the seasonal transformations of society. For all progress, all movement was blanketed by the requirements of defence. Nor on the other hand could literature and scientific advancement exercise their potentially great influence; for the developments which they might have set in motion were at every turn stopped by the same obstacle" (Bolgar 1954, 89).

Within the Byzantine theocracy, the Orthodox Church proved an obstacle to the study of science and natural philosophy. From the ninth to fifteenth centuries, the church sought more to discourage than to facilitate the study of Greek science and natural philosophy. Indeed, philosophy and science were always regarded as the handmaidens of theology, an idea that was eventually abandoned in the medieval

West. The Orthodox Church was hostile to the study of secular disciplines for their own sake. Before the church's deadening influence took hold, Greek scholars in the first three centuries of the Roman Empire made remarkable contributions to science. Some of the greatest names in the history of Greek science flourished in this period. Among these are included the greatest astronomer of antiquity, Claudius Ptolemy (c. A.D. 100–c. 170), and the most renowned physician and medical researcher of the ancient world, Galen (c. A.D. 129–c. 200). Other lesser but nonetheless important contributors could also be mentioned. Indeed, until the end of the sixth century, important contributions to natural philosophy were made in the Byzantine Empire by a number of commentators on the works of Aristotle, such as Alexander of Aphrodisias (fl. 2nd–3rd century A.D.), Themistius (c. A.D. 317–c. 388), Simplicius, and most important of all, the Christian neo-Platonist John Philoponus, whose ideas were destined to have a large impact on both Islamic and Latin natural philosophy.

But the level of achievement was seriously affected in A.D. 529, when, on religious grounds, the emperor Justinian ordered the closing of Plato's Academy in Athens, forcing a number of philosophers to depart the Byzantine Empire and move to the East. After that natural philosophy and science played a minor role in Byzantine intellectual life. This is surprising when we realize that, as compared to their contemporary counterparts in Islam and the Latin West, Byzantine scholars were truly fortunate, because their native language was Greek. They could read, study, and interpret, without problems of translation, all the works available in the Greek language that had accumulated in the Byzantine Empire, especially in Constantinople, since the fifth and fourth centuries B.C. Indeed, most of our Greek manuscripts come from Byzantium. And yet, Byzantine scholars appear not to have taken advantage of this readily available treasure house of science and natural philosophy. Although many of the works of Byzantine scholars lie unread in libraries and archives, especially in Istanbul, it is not likely that discoveries of previously unknown works will alter our overall judgment of their scholarly contributions. The explanation lies in the fact that the attitude of Byzantine scholars was overwhelmingly backward looking, as is evident from a statement by Theodore Metochites, a fourteenth-century student of classical thought, who declared in the preface of his *Historical and Philosophical Miscellanies*: "The great men of the past have said everything so perfectly that they have left nothing for us to say" (Runciman 1970, 94). This negative attitude may be compared to Islamic and

Western Christian scholars, who often went beyond the ancient Greek authorities and regarded it as wholly appropriate to disagree with them and thereby add to the sum total of human knowledge. It is a paradox of history that the civilizations of Islam and Western Europe contributed significantly to the store of human knowledge, using translated works and often lacking important earlier texts, while the Byzantines, who had command of the Greek language and easy access to the manuscript sources of their great Greek predecessors, failed to capitalize on their good fortune.

Despite a generally negative assessment of Byzantine contributions, there were periods during the eleventh century, and especially during the fourteenth and fifteenth centuries, even as the empire was disintegrating, that Byzantine intellectual life burgeoned forth to such an extent that scholars have labeled these periods "renaissances." In the first half of the fifteenth century, some Byzantine scholars brought knowledge of Greek and Greek manuscripts to Italy, helping to spark what has been called the Italian Renaissance. Although during these "renaissance" periods we find much greater interest in Greek literature and science, no significant works were composed that had any detectable influence.

Constant warfare undoubtedly sapped the intellectual strength of Byzantine intellectual life. But the Orthodox Church also played an inhibiting role. The church sometimes persecuted those scholars whom it viewed as too drawn to pagan, secular thought. The church recognized that it could not stop the study of traditional Greek secular works, from which it itself drew some benefits. But efforts were made, sometimes unconsciously, to keep Hellenism under control. In the ninth century, when the main secular interest favored science, the church preferred to encourage the formal study of language. In the eleventh century, when a secular revolution with a nationalist bias had made the pagan past momentarily popular, the church took over education on a large scale and introduced techniques of study that left the shell of Atticism without its substance. In the fourteenth century, some of the more daring Hellenists were persecuted, and had it not been for the general collapse, the church would no doubt have tried again to get control of the educational system (Bolgar 1954, 89–90).

We saw that in the Latin West theologians embraced Greek science and natural philosophy to such an extent that we can actually speak of a class of theologian–natural philosophers. Because the theologians embraced the study of natural philosophy as essential for theology, the West was able to institutionalize the study of natural philosophy in the universities, so that students all across Europe were routinely exposed to it, as well as to logic. The centuries-long study of natural philosophy by generation after generation of students in Western Europe established the rationalistic approach to nature that was an indispensable prelude to the advent of early modern science. Nothing like this occurred in the Byzantine Empire, where theologians were indifferent or hostile to the study of a secular subject like natural philosophy, which never became a regular subject of study in the schools of Byzantium.

Although they failed to take advantage of their command of the Greek language and advance the legacy they inherited from one thousand years of Greek science and natural philosophy, Byzantium did make a momentous contribution to the ultimate advancement of science: Byzantine scholars preserved the texts of Greek science and natural philosophy. It was from the Byzantine Empire that the manuscripts of Greek scientific texts were transmitted to the emerging civilizations of Islam and Western Europe, where they were eventually translated into Arabic and Latin. This vital contribution more than makes up for the failure of Byzantine scholars to do intellectual justice to the treasures that lay at their disposal for so many centuries.

ISLAM

If the Byzantines failed to take genuine advantage over their good fortune to possess the Greek manuscript treatises of ancient science and natural philosophy, those to whom the Greek language was a foreign tongue, and who had to read those treatises in translations, were destined to carry on the traditions of ancient science in ways that far surpassed anything achieved by their Byzantine predecessors and contemporaries. Greek science and natural philosophy were translated and made available to the new civilization of Islam, which first appeared in history during the seventh century A.D. In contrast to Christianity, which was disseminated slowly, taking almost 400 years to become the state religion of the Roman Empire, Islam was transmitted with remarkable speed, taking approximately one hundred years to become the dominant religion in a vast geographic area.

The contrasts between the beginnings of Islam and Christianity are striking. Without armies at its disposal, Christianity spread by missionary zeal; Islam spread largely by military conquest. Where Muslim armies prevailed, the Muslim religion was installed. The conversion of the conquered people was encouraged and facilitated. Indeed, the aim of Muslim armies was to convert conquered peoples. Another striking difference lay in the fact that Christianity was born within the Roman Empire and was subordinate to it for four centuries, during which time Christians adjusted to pagan Greek thought and learned to use it for their own purposes. By contrast, Islam was born outside of the Roman Empire and was never in a subordinate position with respect to other religions and governments. Thus, Islam did not have to adjust to one or more alien cultures or to Greek philosophy and science. Although Greek science and natural philosophy eventually played a significant role in Islamic thought, it was always an outside, alien force, as is apparent from the fact that Greek science and natural philosophy were known within Islam as the "foreign sciences," in contrast to the "Islamic sciences" that were based on the Qur'an and Islamic law and traditions, which always held the highest place in Islamic life.

Despite the fact that the "foreign sciences" were not part of the initial Islamic tradition, they were destined to become part of it as Islam moved onto the world stage. Muslim scholars, and a number of Christians and Jews who lived within the civilization of Islam, came to look with favor on the idea of absorbing the fruits of Greek science and natural philosophy. To accomplish this, it was necessary to translate much of the Greek legacy in science and natural philosophy into Arabic, the language of the Qur'an and of Muslims in the heartland of Islam in the Middle East. The translating activity began even before Islam was born. Christian scholars in Syria and Persia, many of whom were native Greek speakers, began translating Greek texts into Syriac, a Semitic language related to Arabic. During the ninth and tenth centuries, when the Arabic phase of translations was underway, numerous scholars knowledgeable in both Greek and Arabic, or Syriac and Arabic, translated many of the great Greek treatises into Arabic. So intense was the desire for Greek science and natural philosophy, that A. I. Sabra, the eminent historian of Islamic science, has characterized the massive translating activity of the ninth and tenth centuries as an "appropriation" rather than a mere passive "reception" (Sabra 1987, 226–229). These translations laid the foundation for Islam's great contributions to science and natural philosophy over the next four or five centuries.

Islamic contributions to the exact sciences and medicine occasioned no concern, because those disciplines were not controversial and were indeed regarded as useful. Great contributions were made in a variety of sciences by al-Biruni (A.D. 973–d. after 1050), perhaps the most far-ranging scientific writer of the Islamic world, who left treatises on astronomy, geography, chronology, mathematics, mechanics, medicine, and pharmacology, as well as other fields. The list of Islamic scientists and natural philosophers to the beginning of the thirteenth century is lengthy and illustrious. Among the most noteworthy are al-Khwarizmi (fl. c. A.D. 800–847; algebra and arithmetic); al-Farghani (d. after A.D. 861; astronomy); al-Kindi (c. A.D. 801–c. 866; optics, medicine, music, and natural philosophy), known as the "first Arab philosopher"; al-Battani (fl. c. A.D. 880; astronomy); Ibn al-Haytham (A.D. 965–c. 1040; optics, astronomy, and mathematics), known in the West as Alhazen; Omar Khayyam (c. A.D. 1048–c. 1131; various aspects of mathematics, especially algebra in which he solved cubic equations; also a famous poet best known in the West for his *Rubaiyat*); al-Khazini (fl. c. A.D. 1115–c. 1130; astronomy, mechanics, and scientific instruments); al-Bitruji (c. A.D. 1190; astronomy and natural philosophy), known as Alpetragius in the West.

There were also a number of famous physicians, some of whom exerted a large influence on Western medicine. Noteworthy in this group are al-Razi (c. A.D. 854–c. 935; known in the West as Rhazes, he wrote on measles and smallpox; a number of his works were translated into Latin and were quite influential); Ibn Sina (A.D. 980–1037; known in the West as Avicenna; his famous medical treatise was translated into Latin and used as a textbook in medical schools under the title *Canon of Medicine*); Ibn Rushd (A.D. 1126–1198; known in the West as Averroes, he was not only a physician but, like his predecessor, Ibn Sina, a famous commentator on the works of Aristotle); and Ibn al-Nafis (d. A.D. 1288; medicine; he discovered the lesser, or pulmonary, blood circulation). Many more names could be added to the list of Islamic scientists who contributed significantly to the exact sciences and medicine. These scientific and medical treatises did not stir any animosity or religious hostility. It was quite otherwise with natural philosophy.

In Islamic natural philosophy, Aristotle was the major influence, with Plato playing an indirect role to the extent that his ideas were incorporated into neo-Platonic treatises that were translated into Arabic. Plato's dialogues were too difficult to translate into Arabic and too difficult to utilize. Although Aristotle's treatises were hardly easy to comprehend, they were far more intelligible than Plato's dialogues.

With all of Aristotle's works—except the *Politics*—available in Arabic by the mid-eleventh century, the translating movement that had begun in Baghdad under the Abbasid Caliph al-Mansur (ruled A.D. 754–775) in the eighth century was essentially concluded (Peters 1968, 58–61). Included in what might be called the Islamic Aristotelian corpus were works by Greek commentators of the fourth to sixth centuries, such as John Philoponus and Simplicius. With Aristotle's texts available, and those of some of his late Greek commentators, scholars in the Islamic world began to write their own commentaries as well as independent works utilizing Aristotle's natural philosophy. Among the most important Islamic natural philosophers were al-Kindi, al-Farabi (c. A.D. 870–950), Ibn Sina (see Figure 8.1), Ibn Bajja (d. A.D. 1139), and Ibn Rushd. Some of the works of each of these five natural philosophers were translated into Latin and had a significant impact on Western thought. Indeed, Ibn Rushd had a great impact on Western thought but virtually none in the Islamic world, where he was little known. But there were many other Islamic natural philosophers who were unknown in the West but quite influential in the Islamic world. In Baghdad, the intellectual center of the Islamic world, numerous other natural philosophers were important in the late tenth and eleventh centuries. Moreover, there was an important tradition of philosopher-physicians in Baghdad and elsewhere. One of the greatest was Hunayn ibn Ishaq (A.D. 808–873), an important translator of Aristotle's works and also an author of commentaries and original works in philosophy, natural philosophy, and medicine. The tradition of philosopher-physicians extends to the eleventh and twelfth centuries, with Ibn Sina in the former century, and Ibn Rushd in the latter.

From the ninth to the end of the twelfth centuries, and perhaps even a century or two later, natural philosophy was a vibrant subject discussed and taught by numerous individuals, with a few centers where it was especially concentrated—often at the court of a ruling caliph. If it was vibrant during this period, its existence was also ephemeral and precarious. Throughout the history of medieval Islam, the role of Greek philosophy was problematic. At any particular time, there were those who viewed it favorably, while others, undoubtedly a considerable majority, viewed it, at best, with indifference, and often enough with some degree of hostility. Occasionally, the attitude of a caliph was instrumental in altering attitudes toward natural philosophy, but more often attitudes toward natural philosophy and Greek thought

Figure 8.1. Avicenna (Ibn Sina). (The National Library of Medicine.)

were governed by Muslim religious leaders, who exercised great influence in particular regions or cities. Not only was Greek philosophy
regarded as a foreign science, but the term philosopher (*faylasuf*, or
falasifa) was often employed pejoratively.

In the intellectual hierarchy of medieval Islamic society, scholars distinguish three levels (Huff 1993, 69). Because Islam was a nomocracy, the first level was comprised of legal scholars. Religious law and traditions were valued above all else and, therefore, valued even more than theology. Next in order came the *mutakallimun*, scholars who used Greek philosophy to interpret and defend the Muslim religion. The mutakallimun emphasized rational discourse, to which they added the authority of revelation. And finally, at the bottom were the *falasifa*, the Islamic philosophers, who followed rational Greek thought, especially the thought of Aristotle. Not surprisingly, the philosophers placed greatest reliance on reasoned argument while downplaying revelation. The philosophers sought to develop natural philosophy in an Islamic environment, and, as A. I. Sabra has put it, did so, "often in the face of suspicion and opposition from certain quarters in Islamic society" (Sabra 1994, 3).

Of the three Islamic groups just distinguished—namely, legal scholars, who were almost always traditionalists, the mutakallimun, and philosophers—the traditionalists made no real use of Greek philosophy, largely because they found it a threat to revealed truth and the Islamic faith. In their bitter struggle with each other and with the traditionalists, the mutakallimun and the philosophers made much use of Greek philosophy. The mutakallimun were primarily concerned with the *kalam*, which, according to A. I. Sabra, is "an inquiry into God, and into the world as God's creation, and into man as the special creature placed by God in the world under obligation to his creator" (Sabra 1994, 5). Thus, kalam is a theology that used Greek philosophical ideas to explicate and defend the Islamic faith.

Two groups of mutakallimun have been identified: the Mu'tazilites, who were the more extreme, and the Ash'arites (see Hyman and Walsh 1973, 205). Both groups shared an attitude "against the passive acceptance of authority in matters of faith." It was their intention to replace the "passive acceptance of authority" with "a state of knowledge (*'ilm*) rooted in reason" (Sabra 1994, 9). The Mu'tazilites were regarded as Islamic rationalists who equated the power of reason with that of revelation (Huff 1993, 111). They are said to have "made an outstanding contribution to Islamic thought by the assimilation of a large number of Greek ideas and methods of argument" (Watt 1985, 54). These arguments and methods were not adopted for their own sake but rather for their utility in understanding the Islamic religion. In the ninth century, the Mu'tazilites gained the support of caliphs like al-Mamun and Mutassim, as well as influential intellectuals. The supportive caliphs persecuted those who opposed the Mu'tazilite belief that the Qur'an was created by God. They implemented a virtual inquisition. Because many thought the rationalism of the Mu'tazilites was extreme, Sunni Muslims often regarded them as heretics (Watt 1985, 55). Their ascendancy ended with the rule of the Sunni caliph al-Mutawwakil, who destroyed their movement (see Hoodbhoy 1991, 99–100).

The Ash'arites, who followed the teaching of al-Ash'ari (d. A.D. 935), are the second group of mutkallimun. They broke with Mu'tazilism and replaced it as the main representatives of kalam. Ash'arism, however, was a complicated movement, with some of its followers emphasizing rationalism, while others argued in the traditionalist mode (see Makdisi 1962, 37–80; 1963, 19–39). Although both Mu'tazilites and Ash'arites were severe critics of the philosophers, they were themselves regarded as too rational and were bitterly opposed by more conservative Muslims, both from the Sunni and Shiite sides.

In treating the attitudes toward natural philosophy and science in medieval Islam, it is essential to have a good sense of the relationship between Muslim traditionalism and Muslim rationalism, which were engaged in an ongoing and bitter struggle about the role of Islam in intellectual life. George Makdisi provides a useful way to distinguish between Muslim traditionalism and Muslim rationalism:

The traditionalists made use of reason in order to understand what they considered as the legitimate sources of theology: scripture and tradition. What they could not understand they left as it stood in the sources; they did not make use of reason to interpret the sources metaphorically. On the other hand, the rationalists advocated the use of reason on scripture and tradition; and all that they deemed to contradict the dictates of reason they interpreted metaphorically in order to bring it into harmony with reason. (Makdisi 1963, 22)

The antithetical approaches of the Muslim traditionalists and the Muslim rationalists can be illustrated directly from the mutakallimun themselves, namely from the Mu'tazilites and Ash'arites. What was one to make of anthropomorphic statements in the Qur'an that speak of "the face of Allah, His eyes and hands, his sitting on His throne, and His being seen by the Faithful in Paradise" (Arberry 1957, 22)? The strong tendency in Islam was to take such statements literally. Thus al-Ash'ari himself, for whom reason in theology was still important, declared: "We confess that God is firmly seated on His throne.... We confess that God has two hands, without asking how. . . . We confess that God has two eyes, without asking how. . . . We confess that God has a face" (Arberry 1957, 22).

Mu'tazilites, however, viewed these same statements metaphorically: God has no bodily parts; he has no parts or divisions; he is not finite. They also believed that "He cannot be described by any description which can be applied to creatures, in so far as they are created. . . . The senses do not reach Him, nor can man describe Him by analogy. . . . Eyes do not see Him, sight does not reach Him, phantasy cannot conceive Him nor can He be heard by ears" (Arberry 1957, 23). I am unaware of any analogous discussion in the Christian West during the Middle Ages. Medieval Latin theologians regarded anthropomorphic descriptions of God as metaphorical pronouncements.

The Philosophers in Islam

Of the three groups distinguished earlier, the least popular were the philosophers, whom the mutakallimun and conservative Muslims attacked because they used natural philosophy and logic to acquire truth for its own sake, which usually signified that they were ignoring religion. One of the most significant Ash'arite thinkers, the famous al-Ghazali (A.D. 1058–1111), leveled a devastating attack against philosophy. He was fearful of the detrimental effects on the Islamic religion of subjects like natural philosophy, theology (actually metaphysics), logic, and mathematics. In his famous quasi-autobiographical treatise, *Deliverance from Error*, he explains that religion does not require the rejection of natural philosophy, but that there are serious objections to it because nature is completely subject to God, and no part of it can act from its own essence. The implication is obvious: Aristotelian natural philosophy is unacceptable because it assumes that natural objects can act by virtue of their own essences and natures. That is, Aristotle believed in secondary causation—that physical objects are capable of causing effects in other physical objects. Al-Ghazali found mathematics dangerous because it uses clear demonstrations, thus leading the innocent to think that all the philosophical sciences are equally lucid. As al-Ghazali relates, a man will say to himself, "if religion were true, it would not have escaped the notice of these men [that is, the mathematicians] since they are so precise in this science" (translated in Watt 1953, 33). Al-Ghazali explains further that such a man will be so impressed with what he hears about the techniques and demonstrations of the mathematicians that "he draws the conclusion that the truth is the denial and rejection of religion. How many have I seen," al-Ghazali continues, "who err from the truth because of this high opinion of the philosophers and without any other basis" (Watt 1953, 33). Although al-Ghazali allowed that the subject matter of mathematics is not directly relevant to religion, he included the mathematical sciences within the class of philosophical sciences (i.e., mathematics, logic, natural science, theology or metaphysics, politics, and ethics) and concluded that a student who studied these sciences would be "infected with the evil and corruption of the philosophers. Few there are who devote themselves to this study without being stripped of religion and having the bridle of godly fear removed from their heads" (Watt 1953, 34).

In his great philosophical work *The Incoherence of the Philosophers*, al-Ghazali attacks ancient philosophy, especially the views of Aristotle. He does so by describing and criticizing the ideas of al-Farabi and Avicenna, two of the most important Islamic philosophical commentators on Aristotle. After criticizing their opinions on twenty philosophical problems, including the eternality of the world, that God knows only universals and not particulars, and that bodies will not be resurrected after death, al-Ghazali declares: "All these three theories are in violent opposition to Islam. To believe in them is to accuse the prophets of falsehood, and to consider their teachings as a hypocritical misrepresentation designed to appeal to the masses. And this is blatant blasphemy to which no Muslim sect would subscribe" (al-Ghazali 1963, 249).

Al-Ghazali regarded theology and natural philosophy as dangerous to the faith. He had an abiding distrust of philosophers and praised the "unsophisticated masses of men," who "have an instinctive aversion to following the example of misguided genius." Indeed, "their simplicity is nearer to salvation than sterile genius can be" (al-Ghazali 1963, 3). As one of the greatest and most respected thinkers in the history of Islam, al-Ghazali's opinions were not taken lightly.

In light of al-Ghazali's attack on the philosophers, it is not surprising to learn that philosophers were often subject to persecution by religious leaders. Many religious scholars regarded philosophy, logic, and the foreign Greek sciences in general as useless and even ungodly, because they were not directly useful to religion. Indeed, they might even make one disrespectful of religion (Huff 1993, 68). In the thirteenth century, Ibn as-Salah ash-Shahrazuri (d. A.D. 1245), a religious leader in the field of tradition (*hadith*), declared in a *fatwa* that "he who studies or teaches philosophy will be abandoned by God's favor, and Satan will overpower him. What field of learning could be more despicable than one that blinds those who cultivate it and darkens their hearts against the prophetic teaching of Muhammad." Logic was also targeted, because, as Ibn as-Salah put it, "it is a means of access to philosophy. Now the means of access to something bad is also bad" (Goldziher 1981, 205). Ibn as-Salah was not content to confine his hostility to words alone. In a rather chilling passage, he urges vigorous action against students and teachers of philosophy and logic:

Those who think they can occupy themselves with philosophy and logic merely out of personal interest or through belief in its usefulness are betrayed and duped by Satan. It is the duty of the civil authorities to protect Muslims against the evil that such people can cause. Persons of this sort must be removed from the schools and punished for their cultivation of these fields. All those who give evidence of pursuing the teachings of philosophy must be confronted with the following alternatives: either (execution) by the sword or (conversion to) Islam, so that the land may be protected and the traces of those people and their sciences may be eradicated. May God support and expedite it. However, the most important concern at the moment is to identify all of those who pursue philosophy, those who have written about it, have taught it, and to remove them from their positions insofar as they are employed as teachers in schools. (Goldziher 1981, 206)

Although many others shared the attitude of Ibn as-Salah, logic continued to be used as an ancillary subject in scholastic theology (kalam) and in many orthodox religious schools. But there was enough hostility toward philosophy and logic in Islam to prompt philosophers to keep a low profile. Those who taught it did so privately to students who might have sought them out. Following the translations in the early centuries of Islam, Greek philosophy, primarily Aristotle's, received its strongest support from a number of individuals scattered about the Islamic world. As we have already mentioned, al-Kindi, al-Razi, Ibn Sina, and Ibn Rushd were among the greatest Islamic philosophers. All were persecuted to some extent.

Al-Kindi's case reveals important aspects of intellectual life in Islam. The first of the Islamic commentators on Aristotle, al-Kindi was at first favorably received by two caliphs (al-Mamun and al-Mutassim), but his luck ran out with al-Mutawwakil, the Sunni caliph mentioned earlier. According to Pervez Hoodbhoy, "It was not hard for the ulema [religious scholars] to convince the ruler that the philosopher had very dangerous beliefs. Mutawwakil soon ordered the confiscation of the scholar's personal library. . . . But that was not enough. The sixty-year-old Muslim philosopher also received fifty lashes before a large crowd which had assembled. Observers who recorded the event say the crowd roared approval with each stroke" (Hoodbhoy 1991, 111). The other four scholars were also subjected to some degree of persecution, and a number of them had to flee for their safety.

Persecution and harassment of those who advocated the use of reason to explicate revelation are unknown in the medieval Latin West after the mid-twelfth century, when, as we saw in chapter 5, Bernard of Clairvaux and other traditional theologians opposed the application of reason to theology. In his relentless assault on Peter Abelard, Bernard undoubtedly had much in common with Islamic traditionalist theologians. Bernard's hostile attitude lingered on into the first forty years of the thirteenth century, but only at the University of Paris, where church authorities first banned the books of Aristotle from public or private use, then sought unsuccessfully to censor them. By the 1240s, however, Aristotle's books of natural philosophy were taught and read at the University of Paris. Indeed, they had become the core of the curriculum in the arts faculty of that great medieval university (see Grant 1996, 70–80). After the 1240s and for the rest of the Middle Ages, attacks on reason would have been regarded as bizarre and unacceptable. Some theologians were opposed to certain of Aristotle's ideas, but, like Saint Bonaventure, they used Aristotelian natural philosophy and fully recognized that they could not do theology without it. Scholars were sometimes accused of heresy, and occasionally the church tried to curb the excessive use of logic and natural philosophy in theological treatises, but I know of no instance where religious authorities sought to prevent the study of natural philosophy because it threatened religion. Indeed, as time passed, Aristotelian natural philosophy only became more entrenched in the medieval universities. By the time of the Galileo affair in the seventeenth century, the church went to great lengths to defend and protect Aristotle's natural philosophy.

How different it was in Islam, if we judge by a question that Ibn Rushd (Averroes) posed in the twelfth century in his treatise *On the Harmony of Religion and Philosophy* (see Figure 8.2). In this treatise, Ibn Rushd sought to determine "whether the study of philosophy and logic is allowed by the [Islamic] Law, or prohibited, or commanded— either by way of recommendation or as obligatory" (Averroes 1976,

Figure 8.2. Averroes (Ibn Rushd), the Commentator. (Woodcut portrait, Wellcome Library, London.)

44). In the thirteenth century, Ibn as-Salah ash-Shahrazuri, an expert on the tradition of Islam whom we have already met, issued a written reply (*fatwa*) to a question that asked, in Ignaz Goldziher's words, "whether, from the point of view of religious law, it was permissible to study or teach philosophy and logic and further, whether it was permissible to employ the terminology of logic in the elaboration of religious law, and whether political authorities ought to move against a public teacher who used his position to discourse on philosophy and write about it" (Goldziher 1981, 205).

What is remarkable in all this is the fact that, in the twelfth century, Ibn Rushd and, in the thirteenth century, Ibn as-Salah were grappling with the question of whether, from the standpoint of the religious law, it was legitimate to study science, logic, and natural philosophy, even though these disciplines had been readily available in Islam since the ninth century. Ibn Rushd felt compelled to justify their study, while Ibn as-Salah, astonishingly, denied their legitimacy (as we saw earlier in this chapter). I know of no analogous discussions in the late Latin Middle Ages in which any natural philosopher or theologian felt compelled to determine whether the Bible permitted the study of secular subjects. It was simply assumed that it did.

Even so enlightened an author as Ibn Khaldun (A.D. 1332–1406) was hostile to philosophy and philosophers. On the basis of his great *Introduction to History* (*Muqaddimah*), Ibn Khaldun is regarded as the first historian to write a world history. According to Franz Rosenthal: "The *Muqaddimah* was indeed the first large-scale attempt to analyze the group relationships that govern human political and social organization on the basis of environmental and psychological factors" (Rosenthal 1973, 321).

Despite his brilliance as an historian, Ibn Khaldun included a chapter in the *Muqaddimah* titled "A refutation of philosophy. The corruption of the students of philosophy" (Ibn Khaldun 1958, 3:246–258). In this chapter, Ibn Khaldun condemns the opinions of philosophers as wrong and proclaims to his fellow Muslims that "the problems of physics are of no importance for us in our religious affairs or our livelihoods. Therefore, we must leave them alone" (Ibn Khaldun 1958, 3:251–252). He regarded the study of logic as dangerous to the faithful unless they were deeply immersed in the Qur'an and the Muslim religious sciences to fortify themselves against its methods.

When religious authorities in a society are fearful of the effects of natural philosophy on religious beliefs, and are also sufficiently influential to curtail and weaken the impact of science and natural philosophy on the faithful, they will almost certainly use their powers to limit the spread and dissemination of those secular disciplines. This scenario is avoidable only if at least three conditions exist: (1) Natural philosophy is widely regarded as an independent discipline worthy of study; (2) The state supports and protects natural philosophy; and (3) The religious authorities regard natural philosophy favorably. Without the third condition, it is unlikely that the first two conditions could be attained. In Western Europe during the late Middle Ages, the third condition was clearly in effect, which enabled the first condition to come into being.

None of these conditions were met in Islam, perhaps because Islam is a theocracy in which church and state form a single entity. There is no secular state apparatus distinct from the Islamic religion. As a consequence, the schools, or *madrasas*, had as their primary mission the teaching of the Islamic religion, and paid little attention to the foreign sciences, which, as we saw, were comprised of the science and natural philosophy derived ultimately from the Greeks. The analytic subjects derived from the Greeks certainly did not have equal status with religious and theological subjects. Indeed, the foreign sciences played a rather marginal role in the madrasas, which formed the core of Islamic higher education. Only those subjects that illuminated the Qur'an or the religious law were taught. One such subject was logic, which was found useful not only in semantics but was also regarded as helpful in avoiding simple errors of inference. The primary function of the madrasas, however, was "to preserve learning and defend orthodoxy" (Mottahedeh 1985, 91). In Islam, most theologians did not regard natural philosophy as a subject helpful to a better understanding of religion. On the contrary, it was usually viewed as a subject capable of subverting the Islamic religion and, therefore, as potentially dangerous to the faith. Natural philosophy always remained a peripheral discipline in the lands of Islam and was never institutionalized within the educational system, as it was in Latin Christendom. Hence, it was never able to create a large body of students who would use the techniques and methods of natural philosophy to approach nature and its operations in a wholly rationalistic manner. The absence of a large body of students trained in natural philosophy may well have affected the exact sciences, which eventually faltered and faded. Many of the problems of the exact sciences are drawn from natural philosophy. Without a vibrant, inquisitive natural philosophy that has substantial societal support and encouragement, the exact sciences are not likely to receive the requisite degree of intellectual stimulation to make dramatic advances. They are likely to stagnate and eventually grind to a halt. This is perhaps why the great initial promise of Islamic science and natural philosophy failed to come to fruition, so that between 1500 and 1600, the science and natural philosophy of Western Europe surpassed that of Islam.

THE LATIN WEST

Prior to this chapter, I focused on the relations between science and religion in the Latin West, and it is therefore unnecessary to summarize what has already been said. Instead, I shall compare the natural philosophy in the West to its fate in the Byzantine Empire and Islam, concentrating much more on the latter than the former.

Because the three civilizations with which we are concerned were intensely religious, the attitude of the religious authorities toward a secular natural philosophy that was derived essentially from pagan Greek sources is a vital consideration. If that attitude was sufficiently fearful, and even hostile, it is not likely that natural philosophy could have flourished. Although clerics in all three civilizations had problems with Aristotle's natural philosophy, those in the West eventually embraced it with a zeal and enthusiasm that was truly astonishing. By the end of the thirteenth century, natural philosophy and logic served as the basic curriculum in the arts faculties of all medieval universities.

Perhaps the most striking development in the West was the church's eager acceptance of Greco–Arabic (or Greco–Islamic) natural philosophy and science. The path to this acceptance had been prepared over many centuries, as Christians first arrived at a state in which they were willing to accept secular pagan learning as the handmaiden to theology and it was understood that a good Christian would study such subjects only to the extent that they shed light on the Christian faith. When Aristotle's natural philosophy reached the West in the twelfth and thirteenth centuries, some, if not many, Christian theologians were alarmed at certain of Aristotle's opinions about the physical world, and they sought to ban and then censor his views. By the end of the thirteenth century, however, Aristotle's natural philosophy was no longer a contentious issue. It had been so thoroughly embraced that it formed the basic curriculum for all students in the faculties of arts of medieval universities. This was a momentous achievement. It signified that the Catholic Church and its theologians had fully embraced and accepted Greco–Arabic science and natural philosophy. Without this acceptance, natural philosophy could not have become the basis for a liberal arts education in medieval universities and would therefore not have been institutionalized throughout Western Europe.

It is important to point out that not only did university-trained theologians fully accept and embrace the discipline of natural philosophy, but many, if not most, of them were eager and active contributors to the literature of natural philosophy. It is for that reason that it is wholly appropriate to call them "theologian–natural philosophers." They were equally at home in both disciplines and were keen to import as much natural philosophy as they could into the resolution of theological problems, while avoiding any temptations to theologize natural philosophy. This explains why some medieval theologians can be equated with the best of the secular natural philosophers, such as John Buridan and Albert of Saxony. Some theologians, such as Albertus Magnus and Nicole Oresme, were clearly superior to them.

By their actions, theologians in the West were full participants in the development and dissemination of natural philosophy. They made it possible for the institutionalization of natural philosophy in the universities of the late Middle Ages, and therefore its extensive dissemination. Nothing like this occurred in the Byzantine Empire or in Islam. We saw that in Islam there was often overt hostility to natural philosophy and to natural philosophers, who were derisively called "philosophers."

Why did theologians in the West embrace natural philosophy and logic so ardently? There can be little doubt that they were convinced that these disciplines were essential for the analysis and explication of theology. This attitude was already embodied in the conception of natural philosophy as the handmaiden of theology, an attitude that was formed in the early centuries of Christianity. Over the centuries, theologians had gained much experience in dealing with problematic issues in natural philosophy that were potentially subversive of church doctrine. In the final analysis, the benefits the theologians thought they could derive from the study and use of natural philosophy in theology, and for its own sake, far outweighed any feelings of uneasiness they may have felt.

With regard to natural philosophy, which, as I have argued, was the vital element in preparing the way for early modern science, the attitude of the theologians and of the church they served was instrumental, firstly, in permitting natural philosophy to develop as it did, and secondly, in contributing significantly to the rationalistic and analytic nature of medieval natural philosophy. There was nothing like this among the theologians and clerics of medieval Byzantium and Islam. Indeed, it is very likely their basic hostility toward the claims of natural philosophy that made it, at best, a peripheral activity. To flourish, and take deep and extensive root, natural philosophy requires a tolerant attitude from the society within which it functions. There has to be a strong sense that natural philosophy provides the key to an understanding of the operations of the natural world. In Islam, there was no such confidence in the powers of natural philosophy. It was far better, many of the powerful theologians and opinion makers believed, to trust in the Qur'an and the religious law and traditions to understand and cope with the world. To trust in the reasoning of natural philosophers—the *falasifa*—was regarded by many Muslim clerics as dangerous. As F. E. Peters explains, "when confronted with radical *falsafah*, Islamic orthodoxy reacted with determination and frequently with violence" (Peters 1968, 220).

To conclude this comparison of the relations between science and religion in the three great medieval civilizations, I have left the most profound difference for last—namely, the separation of church and state (see chapter 4 for a brief discussion). The manner in which Christianity developed proved favorable to the theoretical separation of church and state. As a minority religion within the vast Roman Empire, Christians wanted freedom to worship, and if this were granted, they were quite prepared to be good citizens of the Roman state. The Bible itself offered strong support for separation. In his famous response to the query by the Pharisees as to whether his followers should pay taxes to the Roman emperor, Jesus urged that they "Render therefore unto Caesar the things which are Caesar's; and unto God the things that are God's" (Matt. 22:21). Thus did Jesus acknowledge the state and urge his followers to be good citizens. A Christian had two allegiances: one to God and the church, and the other to the secular state. Although church and state within Christendom were often in conflict, with one seeking to dominate the other when the opportunity presented itself, each nonetheless recognized the other as an independent entity. The long Christian experience of living within the Roman Empire, and the absence of explicit biblical support for a theocratic state, made the establishment of a theocracy by the Christian church unlikely. Ironically, a theocracy developed within the Greek Orthodox Church of the Byzantine Empire. This state of affairs was not the doing of the church but rather of the Roman Emperors, who made the eastern Christian church subordinate to their rule and produced a variant version of a theocracy known as Caesaropapism, in which the emperor was head of both church and state. Within Islam, the caliphs were regarded as heads of both church and state.

The Byzantine Empire and Islam paid a heavy price for failing to separate church and state. In both societies, Aristotle's natural philosophy was regarded as potentially dangerous because it encompassed ideas and concepts that were hostile to both religions, and because it was often felt that scholars who focused too much on natural philosophy would either neglect religion or come to regard it as inferior to natural philosophy. Islam's failure to separate church and state nullified an institutional advantage it had over Western Christendom. Where the latter was organized as a centralized, hierarchical religion with a single individual—the Pope—holding ultimate power, Islam was a decentralized religion with no hierarchical structure. What power there was derived from local religious leaders who drew on the support of their fellow Muslims. Under these circumstances, we might expect that freedom of inquiry and the cultivation of a vibrant, sustained natural philosophy would have been more likely to occur within the decentralized Muslim religion than within the highly centralized Catholic Church of Western Europe. As we now know, the reverse occurred: the West developed a lively natural philosophy, whereas in Islam natural philosophy became a peripheral and suspect discipline, whose study could even prove dangerous.

The separation of church and state in Western Europe, however, proved an enormous boon to the development of science and natural philosophy. The church did not view natural philosophy as a discipline that had to be theologized or made to agree with the Bible. Although the church felt threatened in varying degrees by Aristotle's natural philosophy in the first three-quarters of the thirteenth century, by the end of that century Aristotle's natural philosophy was fully accepted by all, and it formed the basis of a university education. Although the Byzantine Empire fell to the Ottoman Turks in 1453, Western Christianity continued on with its separation of church and state. Indeed, as the centuries passed, some nation states of Europe became as powerful as, if not more powerful than, the church, which had a diminished capacity for influencing science and natural philosophy. With the advent of the Protestant Reformation, the sphere of influence of the Catholic Church diminished further. As the nation states gained parity with the church and then surpassed it in power and influence, science and natural philosophy had as much, if not more, to fear from the state as from the church. Although this was not a problem during the late Middle Ages, it became one in the twentieth century, as Nazi Germany and the Soviet Union bear witness. With regard to these two European states, the traditional problems between science and religion became problems between science and the state.

The separation of church and state that was an integral part of Western Christianity from its outset was of momentous significance. It made numerous institutional developments feasible that might not otherwise have occurred. Indeed, the very separation of natural philosophy into the faculty of arts and the location of theology in a separate faculty of theology reveals an understanding that these are different subject areas that require very different treatment. The greatest benefit for science and religion is that each was left relatively free to develop independently of the other, although every individual scientist or theologian was free to incorporate ideas and concepts from the one area into the other. Above all, however, the fear and uncertainty that afflicted all too many Islamic and Byzantine natural philosophers was largely absent in the West. The separation of church and state, and the analogous disciplinary distinction between theology and natural philosophy, made possible the independent development of each of these two fundamental disciplines. Indeed, it is ironic that natural philosophy permeated theology to so great an extent that it transformed it into an analytic discipline, whereas theology had almost no effect on the development of natural philosophy.

The factors that can produce a separation between church and state in any society and civilization are largely rooted in the religious and cultural forces that shape that society. Those factors were largely absent from Byzantine and Islamic civilizations, but were manifestly present in Western Europe during the Middle Ages and thereafter. Without the separation of church and state, and the developments that proceeded as a consequence, the West would not have produced a deeply rooted natural philosophy that was disseminated through Europe by virtue of an extensive network of universities, which laid the foundation for the great scientific advances made in the sixteenth and seventeenth centuries, advances that have continued to the present day.

Primary Sources

The source readings presented below illustrate some of the major ideas discussed in the book: the handmaiden concept of science and natural philosophy with respect to religion and theology (Reading 1) and beliefs of philosophers that were regarded as errors with respect to the faith (Reading 2). The final four readings concern issues relevant to God's power and the cosmos, namely arguments about the eternity of the world (Readings 3 and 4); and the way medieval natural philosophers discussed hypothetical situations about our world and the possible existence of other worlds (Readings 5 and 6).

— 1 —

Roger Bacon (A.D. 1214–c. 1292)
The "Opus Majus" of Roger Bacon. A translation by Robert Belle Burke. 2 vols. Philadelphia: University of Pennsylvania Press, 1928. Vol. 1, pp. 65–67, 72–74

PHILOSOPHY AND SCIENCE ARE WORTHLESS IN THEMSELVES BUT ARE VALUABLE AS HANDMAIDENS OF RELIGION

In this selection, Roger Bacon forcefully advocates the handmaiden approach that many Christians had adopted. The handmaiden concept was

a compromise between rejection of traditional pagan learning and its un-
qualified acceptance. The handmaiden tradition had its beginning with the
church fathers, most notably Clement of Alexandria, Origen, Saint Basil,
and Saint Augustine. Bacon urges Christians to study science and natural
philosophy solely for their utility in explicating and comprehending scripture and the articles of faith.

CHAPTER XV

HENCE it follows of necessity that we Christians ought to employ
philosophy in divine things, and in matters pertaining to philosophy
to assume many things belonging to theology, so that it is apparent
that there is one wisdom shining in both. The necessity of this I wish
to establish not only on account of the unity of wisdom, but because
of the fact that we must revert below to the lofty expressions relating
to faith and theology, which we find in the books of the philosophers
and in the parts of philosophy: so that it is not strange that in philosophy I should touch upon the most sacred truths, since God has given
to the philosophers many truths of his wisdom. The power of philosophy must be applied to sacred truth as far as we are able, for the ex-
cellence of philosophy does not otherwise shine forth, since
philosophy considered by itself is of no utility. The unbelieving
philosophers have been condemned, and "they knew God, and did
not glorify Him as God, and therefore became fools and perished in
their own thoughts," and therefore philosophy can have no worth except in so far as the wisdom of God required it. For all that is left is
in error and worthless; and for this reason Alpharabius [that is, al-
Farabi, the Islamic natural philosopher] says in his book on Sciences
that an untaught child holds the same position with respect to a very
wise man in philosophy as such a man does toward the revelation of
God's wisdom. Wherefore philosophy by itself is nothing, but it then
receives vigor and dignity when it is worthy to assume the sacred wisdom. Moreover, the study of wisdom can always continue in this life
to increase, because nothing is perfect in human discoveries. Therefore we of a later age should supply what the ancients lacked, because
we have entered into their labors, by which, unless we are dolts, we
can be aroused to better things, since it is most wretched to be always
using old discoveries and never be on the track of new ones, as Boetius
says, and as we proved clearly above in the proper place. Christians

likewise ought to handle all matters with a view to their own profession, which is the wisdom of God, and to complete the paths of the unbelieving philosophers, not only because we are of a later age and ought to add to their works, but that we may compel the wisdom of the philosophers to serve zealously our own. For this the unbelieving philosophers do, compelled by truth itself as far as it was granted them: for they refer all philosophy to the divine wisdom, as is clear from the books of Avicenna on Metaphysics and Morals, and from Alpharabius, Seneca, and Tullius [that is, Cicero], and Aristotle in the Metaphysics and Morals. For they refer all things to God, as an army to its chief, and draw conclusions regarding angels and many other things; since the principal articles of the faith are found in them; for as will be set forth in the morals, they teach that there is a God and that he is one in essence, of infinite power and goodness, triune in persons, Father, Son, and Holy Spirit, who created all things out of nothing; and they touch on many things concerning Jesus Christ and the Blessed Virgin. Likewise also they teach us of Antichrist and the angels and of their protection of men, and of the resurrection of the dead and of future judgment and of the life of future happiness promised by God to those obedient to him, and of the future misery which he purposes to inflict on those who do not keep his commandments. They write also innumerable statements in regard to the dignity of morals, the glory of laws, and concerning a legislator who must receive the law from God by revelation, who is to be a mediator of God and men and a vicar of God on earth, the Lord of the earthly world. When it shall be proved that he has received the law from God, he must be believed in all things to the exclusion of all doubt and hesitation; who must direct the whole race in the worship of God and in the laws of justice and peace, and in the practice of virtues because of the reverence of God and because of future felicity. [We must avail ourselves of their teachings] because they wrote that the worship of idols should be destroyed, and because they prophesied of the time of Christ. From whatever source the philosophers got these statements and similar ones, we find them in their books, as a clear proof will show in what follows, and any one can discover the fact who cares to read through the books of the philosophers. For we cannot doubt that these things were written by them, from whatever source they received them. Nor should we be surprised that philosophers write such statements; for all the philosophers were subsequent to the patriarchs

and prophets, as we brought out above in its proper place, and therefore they read the books of the prophets and patriarchs which are in
the sacred text.

CHAPTER XIX

MOREOVER, all speculative philosophy has moral philosophy for its
end and aim. And since the end imposes a necessity on those things
pertaining to the end, as Aristotle says in the second book of the
Physics, therefore speculative science always aspires to its own end,
and elevates itself to it, and seeks useful paths to this end, and for this
reason speculative philosophy is able to prepare the principles of moral
philosophy. Thus therefore are the two parts of wisdom related among
the unbelieving philosophers; but with Christian students of philosophy moral science apart from other sciences and perfected is theology,
which adds to the greater philosophy of the unbelievers the faith of
Christ and truths which are in their nature divine. And this end has its
own speculation preceding, just as the moral philosophy of the unbelievers has its own. There is therefore the same relation between the
ends in view as between the speculations; but the end, namely, the
Christian law, adds to the law of the philosophers the formulated articles of the faith, by which it completes the law of moral philosophy,
so that there may be one complete law. For the law of Christ takes and
assumes the laws and morals of philosophy, as we are assured by the
sacred writers and in the practice of theology and of the Church. Therefore the speculations of Christians preceding their own law must add
to the speculation of the other law those things which are able to teach
and prove the law of Christ, in order that one complete speculation
may arise, whose beginning must be the speculative philosophy of the
unbelieving philosophers; and the complement of this must be added
to theology in accordance with the peculiar characteristics of the Christian law. And for this reason the complete philosophy among Christians must have a much more profound knowledge of divine things
than it has among the unbelieving philosophers; and for this reason
Christians ought to consider philosophy as if it had just been discovered, so that they might make it suitable for its own end; and therefore many things must be added in the philosophy of the Christians,
which the unbelieving philosophers could not know. And there are reasons of this kind rising in us from the faith and from the authorities of
the law and of the sacred writers, who are acquainted with philoso-

phy, and they can form the common points of complete philosophy and theology. And these are recognized by the fact that they must be common to believers and unbelievers, so that they may be so well known, when they are brought forward and proved, that they cannot be denied by the wise and those instructed in the philosophy of the unbelievers. For the unbelieving philosophers are ignorant of many things at present concerning divine matters, and if these were suitably set before them and proved by the principles of the complete philosophy, that is, by the vivacity of reason, which has its origin in the philosophy of the unbelievers, although completed by the faith of Christ, they would receive it without contradiction and would rejoice in regard to the truth set before them, because they are eager for wisdom and are more studious than Christians. I do not say, however, that any one of the special articles of the Christian faith should be received on trial, but there are many common rational truths, which every wise man would easily accept from another, although he might be ignorant of them himself, as every man studious and desirous of knowledge learns many things from another and receives them by rational arguments, although he was formerly ignorant of them.

Those philosophizing should not be surprised, therefore, if they must needs raise philosophy to the level of divine things and theological truths and the authorities of sacred writers, and employ these freely whenever the occasion arises, and prove them when necessary, and by means of these prove other matters; since without doubt philosophy and theology have much in common. The sacred writers not only speak as theologians, but as philosophers, and frequently introduce philosophical subjects. Therefore Christians desiring to complete philosophy ought in their works not only to collect the statements of the philosophers in regard to divine truths, but should advance far beyond to a point where the power of philosophy as a whole may be complete. And for this reason he who completes philosophy by truths of this kind must not on this account be called a theologian, nor must he transcend the bounds of philosophy; since he can handle freely what is common to philosophy and theology and what must be accepted in common by believers and unbelievers. There are many such matters besides the statements of unbelieving philosophers, which belonging as it were within the limits of philosophy the man philosophizing in the right way should collect, wherever he finds them, and he should assemble them as though they were his own, whether they occur in the books of the sacred writers, or in the books of the philosophers, or in sacred Scripture, or in the histories, or elsewhere. For there is no author who does not besides his main theme introduce incidentally some matters which belong elsewhere; and for this reason there is a linking together of sciences, because each thing in a manner is dependent on another. But every one who handles a subject in the way he should must assign what belongs to it, both what is necessary and what befits its worth; and therefore wherever he finds these things he knows, how to recognize his own, and therefore he must seize them as his own and arrange them in their proper places. For this reason the philosophizing Christian can unite many authorities and various reasons and very many opinions from other writings besides the books of the unbelieving philosophers, provided they belong to philosophy, or are common to it and theology, and must be received in common by unbelievers and believers. If this be not done, there will be no perfecting, but much loss. And not only must this be done to complete philosophy, but because of Christian conscience which must reduce all truth to divine truth that the former may be subject to and serve the latter; also for this reason that the philosophy of the unbelievers is essentially harmful and has no value considered by itself. For philosophy in itself leads to the blindness of hell, and therefore it must be by itself darkness and mist.

— 2 —

Giles of Rome (c. A.D. 1243–1316)
Errores Philosophorum. Critical text with notes and introduction by Josef Koch. English translation by John O. Riedl. Milwaukee, WI: Marquette University Press, 1944, pp. 3, 11–15, 25. Reprinted with permission of Marquette University Press

ERRORS OF THE PHILOSOPHERS: PRELUDE TO THE CONDEMNATION OF 1277

The errors attributed to Aristotle and Averroes by Giles of Rome in 1274 formed the basis for many of the errors condemned at Paris in 1277. Of the six philosophers singled out by Giles, only the "errors" of Aristotle and Averroes are included here.

HERE BEGIN THE ERRORS OF THE PHILOSOPHERS ARISTOTLE, AVERROES, AVICENNA, ALGAZEL, ALKINDI AND MAIMONIDES COLLECTED BY BROTHER GILES OF THE ORDER OF ST. AUGUSTINE

And first comes

CHAPTER I

A Compilation of the Errors of Aristotle.

CHAPTER II

In which the Errors of Aristotle are Restated in Summary.
  The following therefore are all his errors in summary:

1. That motion did not have a beginning.
2. That time is eternal.
3. That the world did not have a beginning.
4. That the heavens were not made.
5. That God could not make another world.
6. That generation and corruption had no beginning and will have no end.
7. That the sun will always cause generation and corruption in the sublunary world.
8. That nothing new can proceed directly from God.
9. That the resurrection of the dead is impossible.
10. That God could not make an accident without a subject.
11. That there is only one substantial form in any composite.
12. That it is impossible to admit a first man or a first rainfall.
13. That two bodies cannot in any way be in the same place.
14. That there are as many angels as there are orbs. Thus it follows that there are only fifty-five or forty-seven.

There were some who wished to justify the Philosopher's view about the eternity of the world. But this justification cannot stand, because he bases himself always upon the abovementioned principle in order to demonstrate philosophical truths. In fact he scarcely ever wrote a book on philosophy in which he did not say something pertaining to this principle.

Again, in addition to the aforementioned errors, some people sought to attribute to him the doctrine that God knows nothing outside of Himself, so that this sublunary world is thus unknown to Him. They take as their reason for this imputation, the statements which are found in book XII of the *Metaphysics* in the chapter 'The opinion of the fathers'. But that they do not understand the Philosopher, and that this is not his intention, is clear from the statements made in the chapter On good fortune, where he says that God of His very nature knows both the past and the future. Other errors also are attributed to him. We are not concerned with them, because they arise from a misunderstanding.

CHAPTER IV

A Compilation of the Errors of Averroes.

CHAPTER V

In which the Abovementioned Errors are Restated in Summary.

All the errors of the Commentator, exclusive of those which are also errors of the Philosopher, are as follows:

1. That no law is true, although it can be useful.
2. That an angel cannot move anything directly except a celestial body.
3. That an angel is pure act.
4. That in no production is the power of the producer the entire explanation of the thing produced.
5. That diverse effects cannot at the same time proceed directly from any agent.
6. That God does not have any providence over individuals.
7. That there is no trinity in God.
8. That God does not know singulars.
9. That some things come to be as a result of an inner determinism in matter independently of any order of divine providence.
10. That the intellective soul is not multiplied with the multiplication of bodies, but is numerically one.
11. That it is the sensitive soul that places man in the species man.
12. That there is no more perfect unity produced by the union of the intellective soul with the body than by the union of the mover of the heavens with the heavens.

SAINTS BONAVENTURE AND THOMAS AQUINAS ON THE ETERNITY OF THE WORLD

— 3 —

Saint Bonaventure (A.D. 1221–1274) In St. Thomas Aquinas, Siger of Brabant, St. Bonaventure *On the Eternity of the World* (*De Aeternitate Mundi*). Translated from the Latin with an introduction by Cyril Vollert, SJ, STD, Lottie H. Kendzierski, Paul M. Byrne. Milwaukee, WI: Marquette University Press, 1964, pp. 105, 107–113. Reprinted with permission of Marquette University Press

WHY THE ETERNITY OF THE WORLD IS LOGICALLY IMPOSSIBLE

The eternity of the world, in which Aristotle firmly believed, was, as we saw, contrary to the Christian faith. In this selection, Bonaventure presents a number of arguments to show that the eternity of the world is absurd and contradictory. Many of his arguments were derived ultimately from John Philoponus, a Christian neo-Platonist in the sixth century. Observe that Bonaventure does not appeal to the Bible or faith; his intent is to refute belief in an eternal world solely by logic and reason.

The question is: Has the world been produced in time or from eternity. That it has not been produced in time is shown.

[Bonaventure now presents six arguments from Aristotle and others who believed in the eternity of the world. These are omitted. He then gives his arguments against Aristotle and in favor of a world created in time from nothing.]

But there are arguments to the contrary, based on *per se* known propositions of reason and philosophy.

1. The first of these is: *It is impossible to add to the infinite*. This is *per se* evident because everything which receives an addition becomes more; "but nothing is more than infinite." If the world lacks a begin-

ning, however, it has had an infinite duration, and consequently there can be no addition to its duration. But this is certainly false because every day a revolution is added to a revolution; therefore, etc. If you were to say that it is infinite in past time and yet is actually finite with respect to the present, which now is, and, accordingly, that it is in this respect, in which it is finite, that the "more" is to be found, it is pointed out to you that, to the contrary, it is in the past that the "more" is to be found. This is an infallible truth: If the world is eternal, then the revolutions of the sun in its orbit are infinite in number. Again, there have necessarily been twelve revolutions of the moon for every one of the sun. Therefore the moon has revolved more times than the sun, and the sun an infinite number of times. Accordingly, that which exceeds the infinite as infinite is discovered. But this is impossible; therefore, etc.

2. The second proposition is: *It is impossible for the infinite in number to be ordered.* For every order flows from a principle toward a mean. Therefore, if there is no first, there is no order; but if the duration of the world or the revolutions of the heaven are infinite, they do not have a first; therefore they do not have an order, and one is not before another. But since this is false, it follows that they have a first. If you say that it is necessary to posit a limit (*statum*) to an ordered series only in the case of things ordered in a causal relation, because among causes there is necessarily a limit, I ask why not in other cases. Moreover, you do not escape in this way. For there has never been a revolution of the heaven without there being a generation of animal from animal. But an animal is certainly related causally to the animal from which it is generated. If, therefore, according to Aristotle and reason it is necessary to posit a limit among those things ordered in a causal relation, then in the generation of animals it is necessary to posit a first animal. And the world has not existed without animals; therefore, etc.

3. The third proposition is: *It is impossible to traverse what Is infinite.* But if the world had no beginning, there has been an infinite number of revolutions; therefore it was impossible for it to have traversed them; therefore impossible for it to have come down to the present. If you say that they (i.e., numerically infinite revolutions) have not been traversed because there has been no first one, or that they well could be traversed in an infinite time, you do not escape in this way. For I shall ask you if any revolution has infinitely preceded today's revolution or none. If none, then all are finitely distant from this present one. Consequently, they are all together finite in number and so have a beginning. If some one is infinitely distant, then I ask whether the revolution immediately following it is infinitely distant. If not, then

neither is the former (infinitely) distant since there is a finite distance between the two of them. But if it (i.e., the one immediately following) is infinitely distant, then I ask in a similar way about the third, the fourth, and so on to infinity. Therefore, one is no more distant than another from this present one, one is not before another, and so they are all simultaneous.

4. The fourth proposition is: *It is impossible for the infinite to be grasped by a finite power*. But if the world had no beginning, then the infinite is grasped by a finite power; therefore, etc. The proof of the major is *per se* evident. The minor is shown as follows. I suppose that God alone is with a power actually infinite and that all other things have limitation. Also I suppose that there has never been a motion of the heaven without there being a created spiritual substance who would either cause or, at least, know it. Further, I also suppose that a spiritual substance forgets nothing. If, therefore, there has been, at the same time as the heaven, any spiritual substance with finite power, there has been no revolution of the heaven which he would not know and which would have been forgotten. Therefore, he is actually knowing all of them and they have been infinite in number. Accordingly, a spiritual substance with finite power is grasping simultaneously an infinite number of things. If you assert that this is not unsuitable because all the revolutions, being of the same species and in every way alike, are known by a single likeness, there is the objection that not only would he have known the rotations, but also their effects as well, and these various and diverse effects are infinite in number. It is clear, therefore, etc.

5. The fifth proposition is: *It is impossible that there be simultaneously an infinite number of things*. But if the world is eternal and without a beginning, then there has been an infinite number of men, since it would not be without there being men—for all things are in a certain way for the sake of man and a man lasts only for a limited length of time. But there have been as many rational souls as there have been men, and so an infinite number of souls. But, since they are incorruptible forms, there are as many souls as there have been; therefore an infinite number of souls exist. If this leads you to say that there has been a transmigration of souls or that there is but the one soul for all men, the first is an error in philosophy, because, as Aristotle holds, "appropriate act is in its own matter." Therefore, the soul, having been the perfection of one, cannot be the perfection of another, even according to Aristotle. The second position is even more erroneous, since much less is it true that there is but the one soul for all.

6. The last argument to this effect is: *It is impossible for that which has being after non-being to have eternal being*, because this implies a contradiction. But the world has being after non-being. Therefore it is impossible that it be eternal. That it has being after non-being is proven as follows: everything whose having of being is totally from another is produced by the latter out of nothing; but the world has its being totally from God; therefore the world is out of nothing. But not out of nothing as a matter (*materialiter*); therefore out of nothing as an origin (*originaliter*). It is evident that everything which is totally produced by something differing in essence has being out of nothing. For what is totally produced is produced in its matter and form. But matter does not have that out of which it would be produced because it is not out of God (*ex Deo*). Clearly, then, it is out of nothing. The minor, viz., that the world is totally produced by God, is evident from the discussion of another question.

— **4** —

Saint Thomas Aquinas (c. A.D. 1224–1274) In St. Thomas Aquinas, Siger of Brabant, St. Bonaventure *On the Eternity of the World* (*De Aeternitate Mundi*). Translated from the Latin with an introduction by Cyril Vollert, SJ, STD, Lottie H. Kendzierski, Paul M. Byrne. Milwaukee, WI: Marquette University Press, 1964, pp. 19–25. Reprinted with permission of Marquette University Press

AN ETERNAL WORLD IS LOGICALLY POSSIBLE

Contrary to Bonaventure's position, Thomas argues that there is no logical contradiction in assuming that God not only created the world, but also made it eternal. This was a rather popular interpretation among scholastic theologians during the late Middle Ages and the Renaissance.

1. If we suppose, in accord with Catholic faith, that the world has not existed from eternity but had a beginning of its duration, the question

arises whether it could have existed forever. In seeking the true solution of this problem, we should start by distinguishing points of agreement with our opponents from points of disagreement.

If the question is phrased in such a way as to inquire whether something besides God could have existed forever, that is, whether a thing could exist even though it was not made by God, we are confronted with an abominable error against faith. More than that, the error is repudiated by philosophers, who avow and demonstrate that nothing at all can exist unless it was caused by Him who supremely and in a uniquely true sense has existence.

However, if we inquire whether something has always existed, understanding that it was caused by God with regard to all the reality found in it, we have to examine whether such a position can be maintained. If we should decide that this is impossible, the reason will be either that God could not make a thing that has always existed, or that the thing could not thus be made, even though God were able to make it. As to the first alternative, all parties are agreed that God could make something that has always existed, because of the fact that His power is infinite.

2. Accordingly our task is to examine whether something that is made could have existed forever. If we reply that this is impossible, our answer is unintelligible except in two senses or because there are two reasons for its truth: either because of the absence of passive potentiality, or because of incompatibility in the concepts involved.

The first sense may be explained as follows. Before an angel has been made, an angel cannot be made, because no passive potentiality is at hand prior to the angel's existence, since the angel is not made of pre-existing matter. Yet God could have made the angel, and could also have caused the angel to be made, because in fact He has made angels and they have been made. Understanding the question in this way, we must simply concede, in accordance with faith, that a thing caused by God cannot have existed forever, because such a position would imply that a passive potentiality has always existed, which is heretical. However, this does not require the conclusion that God cannot bring it about that some being should exist forever.

Taken in the second sense, the argument runs that a thing cannot be so made because the concepts are incompatible, in the same way as affirmation and denial cannot be simultaneously true; yet certain people assert that even this is within God's power. Others contend that not even God could make such a thing, because it is nothing. However, it is clear that He cannot bring this about, because the power by which it is supposed to be effected would be self-destructive. Nevertheless, if it is alleged that God is able to do such things, the position is not heretical, although I think it is false, just as the proposition that a past event did not take place involves a contradiction. Hence Augustine, in his book against Faustus, writes as follows: "Whoever says, 'If God is omnipotent, let Him bring it about that what has been made was not made,' does not perceive that what he really says is this: 'If God is omnipotent, let Him bring it about that what is true is false for the very reason that it is true.'" Still, some great masters have piously asserted that God can cause a past event not to have taken place in the past; and this was not esteemed heretical.

3. We must investigate, therefore, whether these two concepts are logically incompatible, namely, that a thing has been created by God and yet has existed forever. Whatever may be the truth of the matter, no heresy is involved in the contention that God is able to bring it about that something created by Him should always have existed. Nevertheless I believe that, if the concepts were to be found incompatible, this position would be false. However, if there is no contradiction in the concepts, not only is it not false, but it is even possible; to maintain anything else would be erroneous. Since God's omnipotence surpasses all understanding and power, anyone who asserts that something which is intelligible among creatures cannot be made by God, openly disparages God's omnipotence. Nor can anyone appeal to the case of sin; sins, as such, are nothing.

The whole question comes to this, whether the ideas, to be created by God according to a thing's entire substance, and yet to lack a beginning of duration, are mutually repugnant or not. That no contradiction is involved, is shown as follows. A contradiction could arise only because of one of the two ideas or because of both of them together; and in the latter alternative, either because an efficient cause must precede its effect in duration, or because non-existence must precede existence in duration; in fact, this is the reason for saying that what is created by God is made from nothing.

4. Consequently, we must first show that the efficient cause, namely God, need not precede His effect in duration, if that is what He Himself should wish.

In the first place, no cause producing its effect instantaneously need precede its effect in duration. Now God is a cause producing His effect, not by way of motion, but instantaneously. Therefore He need not precede His effect in duration. The major premise is clear from induction, based on all instantaneous changes, such as illumination, and the like. It can also be demonstrated by reasoning. In any instant in which a thing is asserted to exist, the beginning of its action can likewise be asserted, as is evident in all things capable of generation; the very instant in which fire begins to exist, it emits heat. But in instantaneous action, the beginning and the end of the action are simultaneous, or rather are identical, as in all indivisible things. Therefore, at any moment in which there is an agent producing its effect instantaneously, the terminus of its action can be realized. But the terminus of the action is simultaneous with the effect produced. Consequently no intellectual absurdity is implied if we suppose that a cause which produces its effect instantaneously does not precede its effect in duration. There would be such an absurdity in the case of causes that produce their effects by way of motion, because the beginning of motion must precede its end. Since people are accustomed to think of productions that are brought about by way of motion, they do not readily understand that an efficient cause does not have to precede its effect in duration. And that is why many, with their limited experience, attend to only a few aspects, and so are overhasty in airing their views.

This reasoning is not set aside by the observation that God is a cause acting through His will, because the will, too, does not have to precede its effect in duration. The same is true of the person who acts through his will, unless he acts after deliberation. Heaven forbid that we should attribute such a procedure to God!

5. Moreover, the cause which produces the entire substance of a thing is no less able to produce that entire substance than a cause producing a form is in the production of the form; in fact, it is much more powerful, because it does not produce its effect by educing it from the potentiality of matter, as is the case with the agent that produces a form. But some agent that produces only a form can bring it about that the form produced by it exists at the moment the agent itself exists, as is exemplified by the shining sun. With far greater reason, God, who produces the entire substance of a thing, can cause His own effect to exist whenever He Himself exists.

Besides, if at any instant there is a cause with which the effect proceeding from it cannot co-exist at that same instant, the only reason is that some element required for complete causality is missing; for a complete cause and the effect caused exist together. But nothing complete has ever been wanting in God. Therefore an effect caused by Him can exist always, as long as He exists, and so He need not precede it in duration.

Furthermore, the will of a person who exercises his will suffers no loss in power. But all those who undertake to answer the arguments by which Aristotle proves that things have always had existence from God for the reason that the same cause always produces the same effect, say that this consequence would follow if He were not an agent acting by His will. Therefore, although God is acknowledged to be an agent acting by His will, it nevertheless follows that He can bring it about that what is caused by Him should never have been without existence.

And so it is clear that no logical contradiction is involved in the assertion that an agent does not precede its effect in duration. As regards anything that does imply logical contradiction, however, God cannot bring it into being. . . .

8. Thus it is evident that the statement that something was made by God and nevertheless was never without existence, does not involve any logical contradiction. If there were some contradiction, it is surprising that Augustine did not perceive it, as this would have been a most effective way of disproving the eternity of the world; and indeed he brings forward many arguments against the eternity of the world in the eleventh and twelfth books of *De civitate Dei* [*The City of God*]; yet he completely ignores this line of argumentation. In fact, he seems to suggest that no logical contradiction is discernible here. . . .

9. Another surprising thing is that the best philosophers of nature failed to discern this contradiction. . . .

12. They also bring in arguments which philosophers have touched on, and then undertake to solve them. One among them is fairly difficult; it concerns the infinite number of souls: if the world has existed forever, the number of souls must now be infinite. But this argument is not to the purpose, because God could have made the world without men and souls; or He could have made men at the time He did make them, even though He had made all the rest of the world from eternity. Thus the souls surviving their bodies would not be infinite. Besides, no demonstration has as yet been forthcoming that God cannot produce a multitude that is actually infinite.

There are other arguments which I forbear to answer at the present time. A reply has been made to them in other works. Besides, some of them are so feeble that their very frailty seems to lend probability to the opposite side.

— 5 —

Albert of Saxony (c. A.D. 1316–1390) *Questions on [Aristotle's] On the Heavens.* Translated by Edward Grant from *Questiones et decisiones physicales insignium virorum: Alberti de Saxonia in octo libros Physicorum; tres libros De celo et mundo; duos libros De generatione et corruptione; . . . Recognitae rursus et emendatae summa accuratione et iudicio Magistri Georgii Lokert Scotia quo sunt tractatus proportionum additi. Paris, 1518, bk. 1, question 9, fols. 93r, col. 2-94r, col. 1*

IS THE WORLD A FINITE OR INFINITE MAGNITUDE?

In questions 6 to 9 of the first book of his *Questions on Aristotle's On the Heavens*, Albert considers whether the world is finite or infinite. He treats a different aspect of the basic question in each of the particular numbered questions. In the ninth question, he inquires whether the world is a finite or infinite magnitude. Substantial segments of the question concern God's supernatural actions whereby he creates or annihilates matter beyond our world and also places a large body in a small space, in a manner analogous to the way the body of Christ exists in the host. Because this is a typical medieval question in natural philosophy, I have used square brackets to illustrate the usual six parts into which a typical question was subdivided (for the six parts, see chapter 6). Any other bracketed text is my addition.

[1] In discussing the finitude or infinitude of the world, we inquire in this fourth [conclusion or question] whether the world is a finite or infinite magnitude.

[2] It seems that it is an infinite magnitude because it occupies the whole [and] such a [thing] is infinite.

Secondly, unless the world is an infinite magnitude, it would seem that an infinite space would exist beyond the world. Thus it follows that there would be an actual infinite, the opposite of which was declared previously. And this is confirmed, because if the world were a finite magnitude, someone at the extremity of the world could extend his hand [beyond], since nothing would be there to impede it. Consequently, a space would seem to be there; but it would not be finite, because there seems no [good] reason why this space should be extended only to a certain point [or magnitude].

Thirdly, unless a place or space existed outside the world, it would follow that the last heaven could not be moved locally, which is false. The falsity of the consequent is obvious since it [i.e., the last heaven] is moved, but not with a species of motion other than local motion. Therefore, the consequence is proved, because then the last heaven could not be in a place and consequently could not change place. But since motion, or local motion, is a change of place, what was said follows, namely that the last heaven is not moved locally.

[3] Aristotle wishes to say the opposite in the text, which he proves as one principal conclusion that the world is a finite magnitude.

[4] Here there will be two articles: the first will be on the principle that is sought. [But] after [it has been shown that] the world is a finite magnitude, the second [article will inquire] whether there is something outside it, say an infinite space or some other thing which has already been touched upon in the argument.

[5] With respect to the first [article], let this be the first conclusion: that the world is a finite magnitude. Let it be proved: according to what was said before, the world is a magnitude. Therefore, it is a finite or infinite magnitude. If a finite magnitude, what has been proposed is had; if [it is] an infinite magnitude, it follows that something would be an actual infinite, the opposite of which was shown in the preceding [discussions].

On the second [article], it seems that there is a space beyond the world, since outside the world God could create a stone. But such [a stone] could not be in something indivisible; indeed, it would seem to occupy some divisible space.

Secondly, because God created such a stone outside the world, he is able to move it rectilinearly and to make it further from the last heaven than before; but he cannot do such things except through a space.

Thirdly, God could create two stones outside the world, one outside the other. Therefore it seems that there is an "outside" beyond the world. But "outside" is [or signifies] a difference of place or space; therefore it seems that there is a space outside the world. This is confirmed, because if such stones were not mutually immediate [or contiguous?], as God [indeed] could create them, then one would seem separated from the other, and consequently outside the heaven there would appear to exist a distance by [means of] which such stones could be created separate [or distant] from each other.

Fourthly, having assumed that God could create two other worlds and that these three worlds would touch, just as spheres are imagined to touch, it would seem that between these three worlds and the points at which they touch there would be an intervening space and an intervening distance; for otherwise, their surfaces would touch, and not their points.

With respect to this [first] article, let this be the first conclusion: there is no body beyond the world, which is proved as follows. Every body or world is part of the world; indeed, not only every body, but every being [is part of the world], since the world is the totality of beings. But neither the world, nor any part of the world, lies outside the world, therefore, etc. Secondly, if a body did exist outside the world, it would be there naturally [or violently]. It cannot be said to be there naturally, because every body, whether simple or mixed, has its natural place elsewhere than outside the world. And since the species of one body does not have several natural places, it follows that such a body could not be there naturally. Nor could it be said to be there violently, for then this place would be natural to some other body, which is impossible, since it is natural neither to a simple or a mixed body.

A second conclusion [is this]: no body can be outside the world naturally. This is proved because no simple or mixed body can be there naturally, because such a body cannot be there naturally, as was already proved, [but also] because, similarly, there cannot be any natural place outside the world. This is proved, for if such a place were natural to some body, it would be a violent place for another [body], which is false, as is obvious from things that have already been proved. And so it is proved that outside the world there is not, nor can there be, a vacuum; for a vacuum is said to be that which is without a body, but where there can be a body. Now although there is no body outside the [last] heaven, [it is also] not [possible] for a body to be there, and consequently no vacuum is there and cannot [possibly] be there. Thus, since the term 'vacuum' is privative, it does not only signify the lack of a body somewhere, but also connotes the aptitude of a body to be there.

A third conclusion [is this]: no motion is able to exist naturally outside the world. This is proved [as follows]: where neither a body nor a place can exist naturally, motion cannot exist. The major [premise] is known from this, namely that every motion requires a body; the minor [premise] is obvious from what has been said.

A fourth conclusion [is this]: outside the world time cannot exist. This is proved [as follows]: where motion cannot exist, there time cannot exist. But, by the preceding conclusion, motion cannot exist outside the world; therefore neither can time. The major [premise] is obvious from this, [namely] that time is the same as motion [i.e., it is the measure of motion], as is said in the fourth [book] of the *Physics*.

A fifth conclusion [is this]: God and the intelligences are neither outside the world, nor inside the world circumscriptively [that is, they do not occupy three-dimensional places or spaces]. This is obvious because such [beings] lack position [or location] and magnitude but are outside the world privatively, that is, they do not have a position in the world, which agrees with [the idea] that God and the intelligences are part of the world although they do not have a position [or location] in the world.

From all these things [i.e., from the preceding conclusions], we conclude what Aristotle concluded in the first book of this [treatise], [namely] that outside the world there is neither body, nor place, nor vacuum, nor motion, nor time. But that outside the world there are alterable beings leading the best lives, taking 'outside' in the privative sense understood in the preceding conclusion.

Furthermore, in order to save [the idea] that, beyond the world, God could create supernaturally one or two stones, or another world, or several worlds, it is not necessary to assume that there is actually a space, or vacuum, or distance, or any such thing beyond the world. This is proved [as follows]: for if there were such things beyond the world, God could annihilate them; but with them annihilated, He could [still] create beyond the world things that He wished, just as now. Therefore [etc.]. Next, I concede that God is able to move such a stone created beyond the world, not with a local motion but with a motion having the same structure [or nature] as a local motion, without, [however,] being nearer or farther from the sides [or surfaces] of the [last] heaven.

According to this conception, we can imagine further that just as God could annihilate all things within the sides [or surfaces] of the heaven without those sides touching [or coming together], so also could He place a very large body, one that is as great as the world, in a very small place, as in a millet seed; [and He could do this] without any condensation or rarefaction, or penetration of the bodies. Just in this way must the body of Christ be imagined [to be] in the small host; for in this small host, the body of Christ is just as great as it is without any condensation and penetration whatever, just as he was suspended on the cross for our sins.

[Furthermore,] according to the same conception, we can imagine that in such a millet seed, a space of 100 leagues, or 1,000, or however many [leagues] we wish, could be created. And we could then imagine that a man [placed] in a millet seed could walk through a thousand leagues from one extreme [of that millet seed] to the other; and many other similar things must be conceded according to this [conception].

[6] To the [principal] arguments, [I respond as follows].

To the first, [I reply that] "the world can occupy the whole" can be [understood as signifying] that the world does not occupy a space other than itself; and thus, just as it does not occupy itself, so it does not occupy another space. Or, to put it another way, it could be replied as was said before when it was said that an infinite is that which occupies the whole; it is true that the whole occupies a true and imaginable space. But I deny this about the world.

To the second [principal argument, where it says that] unless the world were an infinite magnitude, there would be an infinite space beyond the world, I deny the consequence, since it would not be because of this that a finite or infinite space would lie outside the world. Moreover, from the substance of the argument, I say that outside the world, God could create a finite space to any extent that it pleased Him, without making it an infinite space. As confirmation [of the second principal argument, it is said that] if the world were a finite magnitude, then someone existing at the last surface of the heaven could extend his hand. I deny [this]. And when it was said that there would be no impediment resisting [the hand] beyond the world, this pleases me, but I say further that notwithstanding this, [the hand could not be extended beyond the world] because of the lack of a receptacle, as of a place or space, in which the hand could be received when it is extended.

To the third [principal argument, namely that] if there were no place or space outside the heaven, the last sphere could not be moved locally, I concede this. Nevertheless, it is compatible with this that it could be moved with a motion of the same species as local motion. But enough has been said about this in the books of the *Physics*.

Next, there were other arguments made in connection with the second article of the question in which it was also proved that a space could exist beyond the world. These remain to be solved.

To the first [of these arguments, namely that] beyond the world God could create a stone, [let me say that] this pleases me. But I say further that such a stone would not be in any divisible or indivisible space because it would not be in any place, just as the whole world is not in any place. However, I concede that when the stone is created beyond the world, a space is also created beyond the world [at the same time]. But this space would be nothing other than the created stone [itself].

To the second [of these arguments, namely that] God is able to move a stone created beyond the world by extending its distance from the [last] heaven [or sphere]. I say that this is true. But this could not occur unless He also created a space in which it could be moved and by which it would have a fixed position [with respect] to the center [of the world] or with respect to the heaven or parts of the heaven. But after God would release this stone from a fixed position with respect to the heaven, it would not be near or far from the heaven; but this was discussed in the body of the question.

To the third [of these arguments, namely that] God could create two stones beyond the world with one outside the other, I concede this by taking the said [term] 'beyond' [or 'outside'] negatively [with respect] to the sense that one would be in the other, just as one intelligence is not in another [intelligence]. But I say further that one of these stones cannot be made outside the other positively unless a space or distance were created with them by means of which one of them could be outside the other. And I say further that 'outside' taken negatively is not a difference of place or space; but 'outside' taken positively and affirmatively is [a difference of place or space].

To the fourth [of these arguments], one could reply on the same basis [as in the preceding reply to the third argument].

— 6 —

Nicole Oresme (c. A.D. 1320–1382) Le Livre du ciel et du monde. Edited by Albert D. Menut and Alexander J. Denomy. Translated with an introduction by Albert D. Menut. Madison: University of Wisconsin Press, 1968, pp. 167–179. Reprinted by permission of The University of Wisconsin Press. (For those who wish to examine a reprint of this selection, with the addition of twenty-six footnotes, see Grant 1974, pp. 547–554.)

ARGUMENTS TO DEMONSTRATE THAT, CONTRARY TO ARISTOTLE, THE EXISTENCE OF OTHER WORLDS IS POSSIBLE

At the command of the French king Charles V, in 1377, Nicole Oresme translated Aristotle's cosmological treatise, *On the Heavens*, from Latin into French and added a detailed, section-by-section commentary on Aristotle's text. It is Oresme's last known treatise and is regarded as one of his most brilliant works. In it, Oresme disagrees with Aristotle on numerous issues, including the possibility of other worlds, the theme of the following selection. In his lengthy discussion, Oresme proposes numerous interesting and unusual ways to illustrate the feasibility of the existence of other worlds, even though he was convinced that God had not, and would not, create other worlds.

Now we have finished the chapters in which Aristotle undertook to prove that a plurality of worlds is impossible, and it is good to consider the truth of this matter without considering the authority of any human but only that of pure reason. I say that, for the present, it seems to me that one can imagine the existence of several worlds in three ways. One way is that one world would follow another in succession of time, as certain ancient thinkers held that this world had a beginning because previous to this all was a confused mass without order, form, or shape. Thereafter, by love or concord, this mass was disentangled, formed, and ordered, and thus was the world created. And finally after a long time this world will be destroyed by discord and will return to the same confused mass, and again, through concord, another world will then be made. Such a process will take place in the future an infinite number of times, and it has been thus in the past. But this opinion is not touched upon here and was reproved by Aristotle in several places in his philosophical works. It cannot happen in this way naturally, although God could do it and could have done it in the past by His own omnipotence, or He could annihilate this world and create another thereafter. And, according to St. Jerome, Origen used to say that God will do this innumerable times.

Another speculation can be offered which I should like to toy with as a mental exercise. This is the assumption that at one and the same time one world is inside another so that inside and beneath the circumference of this world there was another world similar but smaller. Although this is not in fact the case, nor is it at all likely, nevertheless, it seems to me that it would not be possible to establish the contrary by logical argument; for the strongest arguments against it would, it seems to me, be the following or similar ones. First, if there were another world inside our world, it would follow that our earth is where it is by constraint, because inside this earth and beneath its circumference toward its center would be another heaven and other elements, etc. Also, the earth of the second world would be absolutely massive and at the center of both worlds; and the earth of our world would be empty and concave and neither the whole earth nor any part of it would be at the center. Thus, since their natural places are different, it follows from what is said in Chapter Seventeen that these two worlds are of different form so that the world beneath us and this our world would be dissimilar, etc. Also, all natural bodies are limited in bigness and smallness, for the size of a man could diminish or grow so much that he would no longer be a man, and the same with all bodies. So, the world we have imagined inside our own world and beneath its circumference would be so small that it would not be a world at all, for our sun would be more than 2,000 times the size of the other and each of our stars would be larger than this imaginary world. To pursue our thought, one could dig in the ground deep enough to reach the earth of the other world beneath ours. This is an untenable absurdity. Also, we should have to posit two Gods, one for each world, etc. Likewise, we might assume another world like our own to exist in the moon or some other star, etc. Or we could imagine another world above and another beneath the one which is under our world, etc. To show that these and similar speculations do not preclude the possibility of such a thing, I will posit, first of all, that every body is divisible into parts themselves endlessly divisible, as appears in Chapter One; and I point out that *large* and *small* are relative, and not absolute, terms used in comparisons. For each body, however small, is large with respect to the thousandth part of itself, and any body whatsoever, however large, would be small with respect to a larger body. Nor does the larger body have more parts than the smaller, for the parts of each are infinite in number. Also from this it follows that, were the world to be made between now and tomorrow 100 or 1,000 times larger or smaller than it is at present, all its parts being enlarged or diminished proportionally, everything would appear tomorrow exactly as now, just as though nothing had been changed. And, if a stone in a quarry had a small opening in it or a concavity full of air, it is not necessary to say that this stone is outside its natural place. Likewise, if there were a concavity the size of an apple full of air at the earth's center, it would not follow that the earth was out of its natural place nor that it was there by violence. Also, if such concavity were to become a bit larger and then still larger until it became very large, we could not place a limit upon this growth at which point one could say the earth would be out of its natural place, precisely because large and small are relative terms, as we have already said. Therefore, for the earth to be in its natural place, it is enough that the center of its weight should be the center of the world, regardless of the concavity inside the earth, provided that it be held firmly together. And this is the answer to the first argument; for, if a world were enclosed within a concavity inside our earth, nevertheless our earth would be in its natural place since the center of the world would be the middle or center of its weight. A propos, I say further that, according to Scripture, water is above the heavens or the firmament; whence the psalm says [Ps. 103:2–3]: Who stretchest out the heavens, etc., Who coverest these heavens with water. And, elsewhere: Bless the Lord, ye waters that are above the heavens [see Dan. 3:60]. And if this water were not heavy in substance if not in fact, then it would not be water. For this reason it is said to be solid and as though frozen or solidified and is called the glacial or the crystalline heavens. Accordingly, this heaven or this water is in its natural place, in spite of the fact that all the other heavenly spheres and elements are enclosed within the concavity of this sphere, for it is solid and the center of its weight is the center of the world. To the second argument I reply that, even if this earth were hollow and concave, nevertheless it would be in the center of the world or worlds, just as though this were its proper place, taking *place* in the sense of the second member of the distinction made in reply to an argument in Chapter Seventeen. From this it appears that our earth and the earth of the other world within it would be in the same place. To the third argument, which stated that all natural bodies are limited in quantity, I say that in this world they are limited to one quantity or size and that in another world they would be fixed at other limits, for large and small, as we have said, are relative terms which do not mean variation or difference in form. Accordingly, we see men—all of the same form—larger in one region and smaller in another. To the fourth argument, where it was stated that one could dig deep enough into the earth, etc., I answer that nature would not permit this, any more than one could naturally approach the sky close enough to touch it. To the fifth argument, regarding the possibility of two Gods, it does not follow; for one sovereign God would govern all such worlds, but it is possible that additional intelligences would move the heavenly bodies of one world and other intelligences the heavens of the other world. To the sixth argument, where it was said that by analogy one could say there is another world inside the moon, and to the seventh, where it was posited that there are several worlds within our own and several outside or beyond which contain it, etc., I say that the contrary cannot be proved by reason nor by evidence from experience, but also I submit that there is no proof from reason or experience or otherwise that such worlds do exist. Therefore, we should not guess nor make a statement that something is thus and so for no reason or cause whatsoever against all appearances; nor should we support an opinion whose contrary is probable; however, it is good to have considered whether such opinion is impossible.

The third manner of speculating about the possibility of several worlds is that one world could be entirely outside the other in an imagined space, as Anaxagoras held. This solitary type of other world is refuted here by Aristotle as impossible. But it seems to me that his arguments are not clearly conclusive, for his first and principal argument states that, if several worlds existed, it would follow that the earth in the other world would tend to be moved to the center of our world and conversely, etc., as he has loosely explained in Chapters Sixteen and Seventeen. To show that this consequence is not necessary, I say in the first place that, although *up* and *down* are said with several meanings, as will be stated in Book II, with respect to the present subject, however, they are used with regard to us, as when we say that one-half or part of the heavens is up above us and the other half is down beneath us. But up and down are used otherwise with respect to heavy and light objects, as when we say the heavy bodies tend downward and the light tend upward. Therefore, I say that up and down in this second usage indicate nothing more than the natural law concerning heavy and light bodies, which is that all the heavy bodies so far as possible are located in the middle of the light bodies without setting up for them any other motionless or natural place. This can be understood from a later statement and from an explanation in the fourth chapter, where it was shown how a portion of air could rise up naturally from the center of the earth to the heavens and could descend naturally from the heavens to the center of the earth. Therefore, I say that a heavy body to which no light body is attached would not move of itself; for in such a place as that in which this heavy body is resting, there would be neither up nor down because, in this case, the natural law stated above would not operate and, consequently, there would not be any up or down in that place. This can be clarified by what Aristotle says in Book Four of the *Physics*, namely, that in a void there is no difference of place with respect to up or down. Therefore, Aristotle says that a body in a vacuum would not move of itself. In the eleventh chapter of this first book it appears, according to Aristotle, that, since nothing is lower than the center of the earth, nothing is or can be higher than the circumference or the concavity of the lunar sphere, the place proper to fire, as we have often said. Thus, taking up in the second sense above, beyond or outside of this circumference or heaven there is no up nor down. From this it follows clearly that, if God in His infinite power created a portion of earth and set it in the heavens where the stars are or beyond the heavens, this earth would have no tendency whatsoever to be moved toward the center of our world. So it appears that the consequence stated above by Aristotle is not necessary. I say, rather, that, if God created another world like our own, the earth and the other elements of this other world would be present there just as they are in our own world. But Aristotle confirms his conclusion by another argument in Chapter Seventeen and it is briefly this: all parts of the earth tend toward a single natural place, one in number; therefore, the earth of the other world would tend toward the center of this world. I answer that this argument has little appearance of truth, considering what is now said and what was said in Chapter Seventeen. For the truth is that in this world a part of the earth does not tend toward one center and another part toward another center, but all heavy bodies in this world tend to be united in one mass such that the center of the weight of this mass is at the center of this world, and all the parts constitute one body, numerically speaking. Therefore, they have one single place. And if some part of the earth in the other world were in this world, it would tend toward the center of this world and become united with the mass, and conversely. But it does not have to follow that the portions of earth or of the heavy bodies of the other world, if it existed, would tend to the center of this world because in their world they would form a single mass possessed of a single place and would be arranged in up and down order, as we have indicated, just like the mass of heavy bodies in this world. And these two bodies or masses would be of one kind, their natural places would be formally identical, and likewise the two worlds. In Chapter Twenty Aristotle mentions another argument from what was said in the *Metaphysics*—namely, that there cannot be more than one God and, therefore, it seems there can be only one world. I reply that if God is infinite in his immensity, and, if several worlds existed, no one of them would be outside Him nor outside His power; but surely other intelligences would exist in one world and others in the other world, as already stated. And my reply to this argument is given more fully in Chapter Twenty. He argues again in Chapters Twenty-two and Twenty-three of which the purport is briefly this: this world is composed of all the matter available for the constitution of a world, and outside this world there can be no body or matter whatsoever. So it is impossible that another world exists. In reply, I say in the first place, that, assuming that all the matter now existing or that has ever existed is comprised in our world, nevertheless, in truth, God could create *ex nihilo* new matter and make another world. But Aristotle would not admit this. Thus, I say, secondly, that, assuming that nothing could be made save from matter already existing and considering the replies we have given to Aristotle's first arguments regarding this problem—arguments whose substance he repeats and employs here in the present case—nonetheless he does not prove that another or more than one world besides our own could not now exist or may not always have existed, just as he states this world of ours to exist without beginning or end. He argues again in Chapter Twenty-four that outside this world there is no place or plenum, no void, and no time; but he proves this statement by saying that outside this world there can be no body, as he has shown by the reasoning above to which I have replied; so it is unnecessary to answer this argument again. But my position could be strengthened or restated otherwise; for, if two worlds existed, one outside the other, there would have to be a vacuum between them for they would be spherical in shape; and it is impossible that anything be void, as Aristotle proves in the fourth book of the *Physics*. It seems to me and I reply that, in the first place, the human mind consents naturally, as it were, to the idea that beyond the heavens and outside the world, which is not infinite, there exists some space whatever it may be, and we cannot easily conceive the contrary. It seems that this is a reasonable opinion, first of all, because, if the farthest heaven on the outer limits of our world were other than spherical in shape and possessed some high elevation on its outer surface similar to an angle or a hump and if it were moved circularly, as it is, this hump would have to pass through space which would be empty—a void—when the hump moved out of it. Now, if we assumed that the outermost heaven was not thus shaped or that nature could not make it thus, nevertheless, it is certainly possible to imagine this and certain that God could bring it about. From the assumption that the sphere of the elements or of all bodies subject to change contained within the arch of the heavens or within the sphere of the moon were destroyed while the heavens remained as they are, it would necessarily follow that in this concavity there would be a great expanse and empty space. Such a situation can surely be imagined and is definitely possible although it could not arise from purely natural causes, as Aristotle shows in his arguments in the fourth book of the *Physics*, which do not settle the matter conclusively, as we can easily see by what is said here. Thus, outside the heavens, then, is an empty incorporeal space quite different from any other plenum or corporeal space, just as the extent of this time called eternity is of a different sort than temporal duration, even if the latter were perpetual, as has been stated earlier in this chapter. Now this space of which we are talking is infinite and indivisible, and is the immensity of God and God Himself, just as the duration of God called eternity is infinite, indivisible, and God Himself, as already stated above. Also, we have already declared in this chapter that, since our thinking cannot exist without the concept of transmutation, we cannot properly comprehend what eternity implies; but, nevertheless, natural reason teaches us that it does exist. In this way the Scriptural passage, Job 26:[7], which speaks about God can be understood: Who stretchest out the north over the empty place. Likewise, since apperception of our understanding depends upon our corporeal senses, we cannot comprehend nor conceive this incorporeal space which exists beyond the heavens. Reason and truth, however, inform us that it exists. Therefore, I conclude that God can and could in His omnipotence make another world besides this one or several like or unlike it. Nor will Aristotle or anyone else be able to prove completely the contrary. But, of course, there has never been nor will there be more than one corporeal world, as was stated above.

Annotated Bibliography

Abelard, Peter. 1979. *A Dialogue of a Philosopher with a Jew and a Christian*. Trans. Pierre J. Payer. Toronto: Pontifical Institute of Mediaeval Studies.

Ambrose. 1961. *Hexameron, Paradise, and Cain and Abel*. Trans. John Savage. New York: Fathers of the Church.

Anselm. 1944. *Proslogium; Monologium; An Appendix in Behalf of the Fool by Gaunilon; and Cur Deus Homo*. Trans. from Latin by Sidney Norton Deane. LaSalle, IL: Open Court.

Arberry, A. J. 1957. *Revelation and Reason in Islam: The Forwood Lectures for 1956 Delivered in the University of Liverpool*. London: George Allen & Unwin.

Aristotle. 1941. *The Basic Works of Aristotle*. Ed. Richard McKeon. New York: Random House.

———. 1984. *The Complete Works of Aristotle. The Revised Oxford Translation*. Ed. Jonathan Barnes. 2 vols. Princeton, NJ: Princeton University Press.
Categories, trans. J. L. Ackrill
De Interpretatione (On Interpretation), trans. J. L. Ackrill
Prior Analytics, trans. A. J. Jenkinson
Posterior Analytics, trans. Jonathan Barnes
Topics, trans. W. A. Pickard-Cambridge
Sophistical Refutations, trans. W. A. Pickard-Cambridge
Physics, trans. R. P. Hardie and R. K. Gaye
On the Heavens, trans. J. L. Stocks
On Generation and Corruption, trans. H. H. Joachim
Meteorology, trans. E. W. Webster

On the Soul, trans. J. A. Smith
Sense and Sensibilia, trans. J. I. Beare
On Memory, trans. J. I. Beare
On Sleep, trans. J. I. Beare
On Dreams, trans. J. I. Beare
On Divination in Sleep, trans. J. I. Beare
On Length and Shortness of Life, trans. G.R.T. Ross
On Youth, Old Age, Life and Death, and Respiration, trans. G.R.T. Ross
History of Animals, trans. d'A.W. Thompson
Parts of Animals, trans. W. Ogle
Movement of Animals, trans. A.S.L. Farquharson
Progression of Animals, trans. A.S.L. Farquharson
Generation of Animals, trans. A. Platt
Metaphysics, trans. W. D. Ross
Nicomachean Ethics, trans. W. D. Ross, revised by J. O. Urmson
Eudemian Ethics, trans. J. Solomon
Politics, trans. B. Jowett
Rhetoric, trans. W. Rhys Roberts
Poetics, trans. I. Bywater

Armstrong, A. H., ed. 1970. *The Cambridge History of Later Greek and Early Medieval Philosophy*. Cambridge: Cambridge University Press. Includes important articles on numerous church fathers (Augustine, Saint Basil, Origen, Clement of Alexandria, and others) and also on major neo-Platonists. Articles from this work are cited elsewhere in this bibliography.

Asztalos, Monika. 1992. "The Faculty of Theology." In Hilde de Ridder-Symoens, ed., *A History of the University in Europe*. Vol. I: *Universities in the Middle Ages*. Cambridge: Cambridge University Press, pp. 409–441.

Augustine. 1982. *St. Augustine The Literal Meaning of Genesis: De Genesi ad litteram*. Ed. and trans. John Hammond Taylor. In Johannes Quasten, Walter J. Burghardt, and Thomas Comerford Lawler, eds., *Ancient Christian Writers: The Works of the Fathers in Translation*. Vols. 41–42. New York: Newman.

————. 1996. *Saint Augustine, Teaching Christianity (De Doctrina Christiana)*. Trans. Edmund Hill, OP. In John E. Rotelle, OSA, ed., *The Works of Saint Augustine. A Translation for the 21st Century*. Vol. 11. Hyde Park, NY: New City Press.

Averroes (Ibn Rushd). 1976. *On the Harmony of Religion and Philosophy*. A translation with introduction and notes, of Ibn Rushd's Kitab fasl al-maqal, with its appendix (Damima) and an extract from Kitab al-kashf 'an manahij al-adilla by George F. Hourani. London: Luzac.

Bacon, Roger. 1928. *The Opus Majus of Roger Bacon.* Trans. Robert Belle Burke. 2 vols. Philadelphia: University of Pennsylvania Press.

Barnes, Jonathan. 1982. *Aristotle.* Oxford: Oxford University Press.

———. 1995. "Life and Work." In Jonathan Barnes, ed., *The Cambridge Companion to Aristotle.* Cambridge: Cambridge University Press, pp. 1–26.

———. 2000. *Aristotle: A Very Short Introduction.* Oxford: Oxford University Press. The content seems identical to Barnes 1982. A relatively painless way to come to grips with the essential features of Aristotle's thought.

Basil. 1963. *Saint Basil Exegetic Homilies.* Trans. Sister Agnes Clare Way. In *The Fathers of the Church: A New Translation.* Washington, DC: Catholic University of America Press.

Berman, Harold. 1983. *Law and Revolution: The Formation of the Western Legal Tradition.* Cambridge, MA: Harvard University Press. An outstanding study of the interrelations between canon law and theology, and between law and science, most of which occurred within the intellectual environment of the medieval universities.

Bernard of Clairvaux. 1953. *The Life and Letters of St. Bernard of Clairvaux.* Trans. Bruno Scott James. London: Burns Oates.

Boethius, Anicius Manlius Severinus. 1973a. *The Consolation of Philosophy.* Trans. S. J. Tester. Cambridge, MA: Harvard University Press.

———. 1973b. *The Theological Tractates.* Trans. H. F. Stewart, E. K. Rand, and S. J. Tester. London: William Heinemann.

Bolgar, R. R. 1954. *The Classical Heritage and Its Beneficiaries.* Cambridge: Cambridge University Press.

Bowen, Francis. 1885. *Modern Philosophy from Descartes to Schopenhauer and Hartmann.* New York: Charles Scribner's Sons.

Brown, Joseph E. 1978. "The Science of Weights." In David C. Lindberg, ed. *Science in the Middle Ages.* Chicago: University of Chicago Press, pp. 179–205.

Bulmer-Thomas, Ivor. 1971. "Euclid." In *Dictionary of Scientific Biography* (hereafter cited as *DSB*), Vol. 4, pp. 414–437.

Buridan, John. 1509. *Acutissimi philosophi reverendi Magistri Johannis Buridani subtilissime questiones super octo Physicorum libros Aristotelis.* Facsimile, 1964.

――――. 1942. *Iohannis Buridani Quaestiones super libris quattuor De caelo et mundo*. Ed. Ernest A. Moody. Cambridge, MA: Mediaeval Academy of America.

Campanus of Novara. 1971. *Campanus of Novara and Medieval Planetary Theory: "Theorica planetarum"*. Ed. Francis S. Benjamin, Jr., and G. J. Toomer. Madison: University of Wisconsin Press.

Chadwick, Henry. 1970a. "The Beginning of Christian Philosophy: Justin: The Gnostics." In A. H. Armstrong, ed., *The Cambridge History of Later Greek and Early Medieval Philosophy*. Cambridge: Cambridge University Press, pp. 158–167.

――――. 1970b. "Origen." In A. H. Armstrong, ed., *The Cambridge History of Later Greek and Early Medieval Philosophy*. Cambridge: Cambridge University Press, pp. 182–192.

――――. 1970c. "Philo." In A. H. Armstrong, ed., *The Cambridge History of Later Greek and Early Medieval Philosophy*. Cambridge: Cambridge University Press, pp. 137–157.

――――. 1981. *Boethius: The Consolations of Music, Logic, Theology, and Philosophy*. Oxford: Clarendon Press.

Clagett, Marshall. 1957. *Greek Science in Antiquity*. London: Abelard-Schuman Ltd. A brief but excellent description of the history of science, beginning with ancient Egypt and Mesopotamia, but focusing primarily on Greek and Roman science from their beginnings to the start of the early Middle Ages.

――――. 1959. *The Science of Mechanics in the Middle Ages*. Madison: University of Wisconsin Press. An outstanding collection of carefully organized selections on the science of mechanics, which embraces statics (Part I), kinematics (Part II), and dynamics (Part III). Many source readings are translated and copiously annotated. Part IV ("The Fate and Scope of Medieval Mechanics") ties together all that has gone before and shows how Galileo and Newton benefited from their medieval predecessors.

――――. 1970. "Archimedes." In *DSB*. Vol. 1, pp. 213–231.

Clement of Alexandria. 1983. "Miscellanies." In *The Ante-Nicene Fathers: Translations of the Writings of the Fathers Down to A.D. 325*. Vol. 2: *Fathers of the Second Century: Hermas, Tatian, Athenagoras, Theophilus, and Clement of Alexandria (Entire)*. Grand Rapids, MI: Wm. B. Eerdmans.

Cobban, Alan B. 1975. *The Medieval Universities: Their Development and Organization*. London: Methuen.

Cohen, Morris R., and I. E. Drabkin. 1948. *A Source Book in Greek Science*. New York: McGraw-Hill. An excellent collection of source readings in English translation ranging over the physical and life sciences.

Cooper, L. 1956. *The Poetics of Aristotle*, Rev. ed. Ithaca, NY: Cornell University Press.

Copleston, Frederick. 1953. *A History of Philosophy*. Vol. 3: *Ockham to Suarez*. Westminster, MD: Newman Press.

———. 1957. *A History of Philosophy*. Vol. 2: *Mediaeval Philosophy Augustine to Scotus*. Westminster, MD: Newman Press.

———. 1960. *A History of Philosophy*. Vol. 1: *Greece and Rome*. Rev. ed. Westminster, MD: Newman Press.

Crombie, A. C. 1959. *Medieval and Early Modern Science*. 2 vols. New York: Doubleday. A still useful and enlightening work. Crombie emphasizes the methodology medieval scholars used in their science and natural philosophy.

Cunningham, Andrew. 1991. "How the Principia Got its Name; or Taking Natural Philosophy Seriously." *History of Science* 29: 377–392.

Dales, Richard C., ed. 1973. *The Scientific Achievement of the Middle Ages*. Philadelphia: University of Pennsylvania Press. In this brief source book, Dales interweaves translated source readings with summaries and paraphrases from reliable modern scholars. Only the physical sciences are included; medicine and biology are omitted.

Dictionary of Scientific Biography (cited as *DSB*). 1970–1980. Charles C. Gillispie, ed. 16 vols. New York: Charles Scribner's Sons. Although the *DSB* treats all of science up to the dates of publication, it is nonetheless a rich source of information about ancient and medieval science, containing approximately 400 biographical articles on the most significant contributors to science in these periods. Names and dates for each biographical entry are given in vol. 16 (1980), which is a comprehensive index to the entire set of volumes.

Dictionary of the Middle Ages (hereafter cited as *DMA*). 1982–1989. Joseph Strayer, ed. 13 vols. New York: Charles Scribner's Sons. Separate articles are devoted to many of the best-known medieval natural philosophers; also included are some thematic articles relevant to science and religion.

Diogenes Laertius. 1950. *Lives of Eminent Philosophers*. Trans. R. D. Hicks. 2 vols. Cambridge, MA: Harvard University Press.

Dod, Bernard G. 1982. "Aristoteles latinus." In Norman Kretzmann, Anthony Kenny, and Jan Pinborg, eds., *The Cambridge History of Later Medieval Philosophy*. Cambridge: Cambridge University Press, pp. 45–79.

Dronke, Peter, ed. 1988. *A History of Twelfth-Century Western Philosophy*. Cambridge: Cambridge University Press. A collection of fine articles on major themes, such as logic and science; also includes articles on the major figures of the twelfth century, including Peter Abelard and William of Conches.

Edelstein, Ludwig. 1943. *The Hippocratic Oath: Text, Translation and Interpretation*. In *Supplement to the Bulletin of the History of Medicine*. No. 1. Baltimore, MD: The Johns Hopkins Press.

Erasmus. 1993. *Praise of Folly*. Trans. Betty Radice. London: Penguin Books.

Euclid. 1956. *The Thirteen Books of Euclid's Elements. Translated from the text of Heiberg, with introduction and commentary by Sir Thomas L. Heath*. 2nd ed. 3 vols. New York: Dover Publications.

Ferngren, Gary B., ed. 2002. *Science and Religion: A Historical Introduction*. Baltimore: Johns Hopkins University Press. Part II ("The Premodern Period") includes four articles about medieval science and religion: "Aristotle and Aristotelianism"; "Early Christian Attitudes Toward Nature"; "Medieval Science and Religion"; and "Islam."

Galileo Galilei. 1962. *Galileo: Dialogue Concerning the Two Chief World Systems*. Trans. Stillman Drake. Berkeley and Los Angeles: University of California Press. In this momentous work, Galileo defended Copernicus' heliocentric system enunciated in 1543 and sought to subvert Aristotle's earth-centered cosmology. It was on the basis of this work, published in 1632, that Galileo was condemned by the Catholic Church in 1633.

Gaybaa, B. P. 1988. *Aspects of the Mediaeval History of Theology 12th to 14th Centuries*. Pretoria: University of South Africa.

Ghazali, al-. 1963. *Al-Ghazali's Tahafut al-Falasifah [Incoherence of the Philosophers]*. Trans. Sabih Ahmad Kamali. Pakistan Philosophical Congress Publication, no. 3.

Giles of Rome. 1944. *Giles of Rome: Errores Philosophorum*. Critical text with notes and introduction by Josef Koch. Trans. John O. Riedl. Milwaukee, WI: Marquette University Press.

Gillispie, Charles C., ed. See *DSB*.

Gilson, Etienne. 1938. *Reason and Revelation in the Middle Ages*. New York: Charles Scribner's Sons.

————. 1955. *History of Christian Philosophy in the Middle Ages*. London: Sheed and Ward. A lengthy history of philosophical thought from the Greek Apologists, who defended the Christian faith, to Nicholas of Cues (or Cusa) in the fifteenth century. The end notes are extensive and detailed. The work is a mine of information.

Goldziher, Ignaz. 1981. "The Attitude of Orthodox Islam Toward the 'Ancient Sciences'." In Merlin Swartz, ed., *Studies on Islam*. New York: Oxford University Press, pp. 185–215.

Gracia, Jorge J. E., and Timothy B. Noone, eds. 2003. *A Companion to Philosophy in the Middle Ages*. Malden, MA: Blackwell Publishing. A splendid resource for the study of medieval philosophy and natural philosophy. Not only does it include a section on the historical context of medieval philosophy, but the main part of the book consists of 138 biographies of the most important philosophers, natural philosophers, and theologian–natural philosophers in the Middle Ages. Each biographical article is accompanied by a bibliography.

Grant, Edward. 1974. *A Source Book in Medieval Science*. Cambridge, MA: Harvard University Press. A heavily annotated, comprehensive source book on all aspects of science and natural philosophy. The work concludes with brief biographies of each author.

————. 1981. *Much Ado About Nothing: Theories of Space and Vacuum from the Middle Ages to the Scientific Revolution*. Cambridge: Cambridge University Press. Describes the significant role played by scholastic natural philosophers and theologians in shaping the spatial conceptions of seventeenth-century scientists. The two major themes concern medieval ideas about intra-cosmic and extra-cosmic void space. The discussion extends to the conceptions of infinite space proposed by non-scholastic authors in the sixteenth and seventeenth centuries, including Henry More, John Locke, and Isaac Newton.

————. 1994. *Planets, Stars, and Orbs: The Medieval Cosmos, 1200–1687*. Cambridge: Cambridge University Press. A comprehensive and detailed study of medieval ideas in two major parts: The first treats the cosmos as a whole and what, if anything, lies beyond it (topics include the creation of the world and various biblical themes relevant to medieval cosmology); the second concerns the celestial region in which some twenty topics are discussed; two extensive appendices and a bibliography conclude the volume.

————. 1996. *The Foundations of Modern Science in the Middle Ages: Their Religious, Institutional, and Intellectual Contexts*. Cambridge: Cambridge University Press. As the title suggests, the primary thesis proclaims that the legitimate pre-history of modern science occurred in the late Middle Ages. Instrumental in producing the intellectual climate that brought forth early modern science were the translations, the universities, and the relations between theology and natural philosophy.

———. 1999. "God, Science, and Natural Philosophy in the Late Middle Ages." In Lodi Nauta and Arjo Vanderjagt, eds., *Between Demonstration and Imagination: Essays in the History of Science and Philosophy Presented to John D. North.* Leiden: Brill, pp. 243–267.

———. 2001. *God and Reason in the Middle Ages.* Cambridge: Cambridge University Press. Describes the emergence of reason and reasoned argument as a powerful instrument in medieval philosophical and scientific discussions. Major chapters are devoted to the role of reason in logic, natural philosophy, and theology as these were taught in medieval universities. The final chapter ("Assault on the Middle Ages") shows the manner in which the Middle Ages came to be falsely viewed as a backward, superstitious period largely devoid of serious intellectual accomplishments.

Guerlac, Rita. 1979. *Juan Luis Vives Against the Pseudodialecticians: A Humanist Attack on Medieval Logic: The Attack on the Pseudodialecticians* and *On Dialectic.* Book III, v, vi, vii from *The Causes of the Corruption of the Arts, with an Appendix of Related Passages by Thomas More.* The texts, with translation, introduction, and notes by Rita Guerlac. Dordrecht, Holland: D. Reidel.

Haskins, Charles Homer. 1957a. *The Renaissance of the Twelfth Century.* Cleveland, OH: World Publishing Co. [Meridian Books]. (Orig. pub. 1927.) Describes the manner in which the twelfth century was a new beginning—a renaissance—for intellectual life in Western Europe.

———. 1957b. *The Rise of Universities.* Ithaca, NY: Great Seal Books, Cornell University Press. (Orig. pub. 1923.) Of the three chapters in this brief but readable book, one is on the medieval professor; another is on the medieval student.

Heath, Sir Thomas. 1921. *A History of Greek Mathematics.* 2 vols. Oxford: Clarendon Press.

Hollister, Warren C., ed. 1969. *The Twelfth Century Renaissance.* New York: Wiley.

Hollister, Warren C. 1994. *Medieval Europe: A Short History.* 7th ed. New York: McGraw-Hill.

Holopainen, Toivo J. 1996. *Dialectic and Theology in the Eleventh Century.* Leiden: E. J. Brill.

Hoodbhoy, Pervez. 1991. *Islam and Science: Religious Orthodoxy and the Battle for Rationality*. London: Zed Books. An unusual book by a physicist in Pakistan who, in seeking to counter anti-scientific tendencies in modern Pakistan and elsewhere in Islam, presents a series of chapters on the history of Islamic science and considers why the scientific revolution did not occur in Islam.

Huff, Toby. 1993. *The Rise of Early Modern Science: Islam, China, and the West*. Cambridge: Cambridge University Press. An excellent study of the development and status of science and natural philosophy in the three civilizations mentioned in the title. Huff presents a detailed comparison between Islam and the West with respect to law, reason and rationality, and education, and provides a sound interpretation as to why modern science emerged in the West and not in Islam.

Hugh of St. Victor. 1961. *The "Didascalicon" of Hugh of St. Victor: A Medieval Guide to the Arts*. Trans. Jerome Taylor. New York: Columbia University Press.

Hyman, Arthur, and James J. Walsh, eds. 1973. *Philosophy in the Middle Ages: The Christian, Islamic, and Jewish Traditions*. Indianapolis: Hackett. A source book of well-chosen readings relevant to philosophy, drawn from major figures in the three monotheistic traditions, beginning with Saint Augustine and concluding with John Buridan. All told, fifteen Christian, four Islamic, and three Jewish authors are represented.

Ibn Khaldun. 1967. *The Muqaddimah: An Introduction to History*. Trans. Franz Rosenthal. 2nd ed. 3 vols. Princeton, NJ: Princeton University Press.

John of Damascus. 1958. *Saint John of Damascus, Writings*. Trans. Frederic H. Chase. New York: Fathers of the Church.

John of Salisbury. 1955. *The "Metalogicon" of John of Salisbury: A Twelfth-Century Defense of the Verbal and Logical Arts of the Trivium*. Trans. Daniel D. McGarry. Berkeley and Los Angeles: University of California Press.

Jones, Charles W. 1970. "Bede, the Venerable." In *DSB*. Vol. 1, pp. 564–566.

Kaiser, Christopher B. 1997. *Creational Theology and the History of Physical Science: The Creationist Tradition from Basil to Bohr*. Leiden: Brill. The first two of five chapters are about the creationist tradition in its various manifestations in late antiquity and the early Middle Ages (chapter 1) and the relations between the church and Aristotelian science, or theology and natural philosophy, in the late Middle Ages (chapter 2). The author, a physicist, has produced a well-written, synthetic account of the relations between science and religion centered on creational issues.

Knowles, David. 1962. *The Evolution of Medieval Thought*. Baltimore: Helicon Press.

Le Goff, Jacques. 1993. *Intellectuals in the Middle Ages*. Trans. Teresa Lavender Fagan. Cambridge, MA: Blackwell Publishers. An overall summary of the broad aspects of medieval intellectual life, but with much emphasis on university education.

Lerner, Ralph, and Muhsin Mahdi, eds. 1963. *Medieval Political Philosophy: A Sourcebook*. Ithaca, NY: Cornell University Press.

Lewes, George Henry. 1864. *Aristotle: A Chapter from the History of Science, including Analyses of Aristotle's Scientific Writings*. London: Smith, Elder and Co.

Lindberg, David C. 1976. *Theories of Vision from al-Kindi to Kepler*. Chicago: University of Chicago Press.

―――. 1983. *Roger Bacon's Philosophy of Nature: A Critical Edition, with English Translation, Introduction, and Notes, of "De multiplicatione specierum" and "De speculis comburentibus"*. Oxford: Clarendon Press.

―――. 1992. *The Beginnings of Western Science: The European Scientific Tradition in Philosophical, Religious, and Institutional Context, 600 B.C. to A.D. 1450*. Chicago: University of Chicago Press. A thorough, very well-written survey of the history of science, it is presently the best study of medieval science available.

Lindberg, David C., ed. 1978. *Science in the Middle Ages*. Chicago: University of Chicago Press. An important collection of articles that encompass not only the sciences and medicine, but also broader aspects involving technology, economic progress, the philosophical environment, and the transmission of science to the West.

Lindsay, Jack. 1971. *Origins of Astrology*. London: Frederick Muller.

Lloyd, G.E.R. 1968. *Aristotle: The Growth and Structure of His Thought*. Cambridge: Cambridge University Press. Intended for beginning students of Aristotle's life and thought. Students will derive much benefit from Lloyd's chapters on the physics of the celestial and sublunary regions.

―――. 1970. *Early Greek Science: Thales to Aristotle*. New York: W. W. Norton.

―――. 1973. *Greek Science After Aristotle*. New York: W. W. Norton. This brief volume and the one preceding form an excellent overall treatment of ancient Greek science from its beginnings in the sixth century B.C. to its end in the sixth century A.D.

Logan, Donald F. 1989. "Vikings." In *DMA*. Vol. 12, pp. 422–437.

Lones, Thomas E. 1912. *Aristotle's Researches in Natural Science*. London: West, Newman, and Co.

Luscombe, D. E. 1988. "Peter Abelard." In Peter Dronke, ed., *A History of Twelfth-Century Western Philosophy*. Cambridge: Cambridge University Press, pp. 279–307.

Luther, Martin. 1957. *Luther's Works*. Vol. 31: *Career of the Reformer*: I, Ed. Harold J. Grimm. Philadelphia: Muhlenberg Press.

Maccagnolo, Enzo. 1988. "David of Dinant and the Beginnings of Aristotelianism in Paris." Trans. Jonathan Hunt. In Peter Dronke, ed., *A History of Twelfth-Century Western Philosophy*. Cambridge: Cambridge University Press, pp. 429–442.

Macrobius. 1952. *Commentary on the Dream of Scipio*. Trans. William Harris Stahl. New York: Columbia University Press.

Makdisi, George. 1962, 1963. "Ash'ari and the Ash'arites in Islamic Religious History," Pts. 1 and 2. *Studia Islamica* 17 (1962): 37–80; 18 (1963): 19–39.

Manchester, William. 1992. *A World Lit Only by Fire: The Medieval Mind and the Renaissance: Portrait of an Age*. Boston: Little, Brown & Co.

Marenbon, John. 1981. *From the Circle of Alcuin to the School of Auxerre: Logic, Theology and Philosophy in the Early Middle Ages*. Cambridge: Cambridge University Press.

Minio-Paluello, Lorenzo. 1973. "James of Venice." In *DSB*. Vol. 7, pp. 65–67.

———. 1974. "William of Moerbeke." In *DSB*. Vol. 9, pp. 434–440.

Montgomery, Scott L. 2000. *Science in Translation: Movements of Knowledge Through Culture and Time*. Chicago: University of Chicago Press. An insightful study of the transmission of scientific knowledge—especially astronomy—by translation. The author describes the numerous problems associated with translation and the vital role it played in the development of science.

Mottahedeh, Roy. 1985. *The Mantle of the Prophet: Religion and Politics in Iran*. New York: Pantheon Books.

Murdoch, John E. 1969. "*Mathesis in Philosophiam Scholasticam Introducta:* The Rise and Development of the Application of Mathematics in Fourteenth-Century Philosophy and Theology." In *Arts Libéraux et Philosophie au Moyen Age; Actes du Quatrième Congrès International de Philosophie Médiévale*. Montreal: Institut d'études médiévales. Paris: Vrin, pp. 215–254.

————. 1975. "From Social into Intellectual Factors: An Aspect of the Unitary Character of Late Medieval Learning." In John E. Murdoch and Edith Dudley Sylla, eds., *The Cultural Context of Medieval Learning*. Dordrecht, Holland: D. Reidel, pp. 271–339.

————. 1984. *Album of Science: Antiquity and the Middle Ages*. New York: Charles Scribner's Sons. An extraordinary collection of illustrations representing many aspects of science and learning drawn from a large number of medieval manuscripts; the explanations and commentaries for each illustration are detailed and illuminating.

Nicholas, David. 1992. *The Evolution of the Medieval World: Society, Government and Thought in Europe, 312–1500*. New York: Longman. A thorough account of medieval development, with a helpful section on education and intellectual life.

Oesterle, J. A. 1967. "Poetics (Aristotelian)." In *New Catholic Encyclopedia*. Vol. 11. Washington, DC: Catholic University of America Press, pp. 455–457.

Oresme, Nicole. 1968. *Nicole Oresme: Le Livre du ciel et du monde*. Ed. Albert D. Menut and Alexander J. Denomy. Trans. Albert D. Menut. Madison, WI: University of Wisconsin Press.

————. 2000. *Nicole Oresme's On Seeing the Stars (De visione stellarum): A Critical Edition of Oresme's Treatise on Optics and Atmospheric Refraction*. Trans. Danny Ethus Burton. PhD diss., Indiana University.

Origen. 1980. "Letter of Origen to Gregory." In Allan Menzies, ed., *The Ante-Nicene Fathers: Translations of the Writings of the Fathers Down to A.D. 325*. 5th ed. Vol. 10. Grand Rapids, MI: Wm. B. Eerdmans.

Peters, F. E. 1968. *Aristotle and the Arabs: The Aristotelian Tradition in Islam*. New York: New York University Press.

Philo of Alexandria (Philo Judaeus). 1929–1962. *Philo with an English translation*. Ed. and trans. F. H. Colson, G. H. Whitaker, and R. Marcus. Vol. 1. Cambridge, MA: Harvard University Press.

Piltz, Anders. 1981. *The World of Medieval Learning*. Trans. David Jones. Totowa, NJ: Barnes & Noble Books. (Orig. pub. 1978.) Although Piltz focuses his attention on Thomas Aquinas, he provides an interesting study of teacher and student life in the Middle Ages and includes some translated source readings, as well as numerous illustrations drawn from early printed editions, each accompanied by a detailed description of its significance.

Plato. 1955. *Plato's "Phaedo": The "Phaedo" of Plato*. Trans. R. S. Bluck. New York: Liberal Arts Press.

———. 1957. *Plato's Cosmology: The "Timaeus" of Plato.* Trans. Francis M. Cornford. New York: Liberal Arts Press.

Ptolemy, Claudius. 1984. *Ptolemy's Almagest.* Trans. G. J. Toomer. New York: Springer-Verlag.

Reid, Jimmy. 1998. "A bid to the oil wheels of a runaway power." *Glasgow Herald*, December 23, 1998.

Ridder-Symoens, Hilde de, ed. 1992. *A History of the University in Europe.* Volume 1: *Universities in the Middle Ages.* Cambridge: Cambridge University Press. A thorough study of the history of medieval universities. All aspects of university life are covered, especially curriculum and faculties (separate chapters are devoted to each of the four faculties: arts, theology, law, and medicine).

Rosenthal, Franz. 1973. "Ibn Khaldun." In *DSB.* Vol. 7, pp. 320–323.

Rüegg, Walter. 1992. "Themes." In Hilde de Ridder-Symoens, ed., *A History of the University in Europe.* Vol. I: *Universities in the Middle Ages.* Cambridge: Cambridge University Press, pp. 3–34.

Rummel, Erika. 1995. *The Humanist-Scholastic Debate in the Renaissance and Reformation.* Cambridge, MA: Harvard University Press.

Runciman, Steven. 1970. *The Last Byzantine Renaissance.* Cambridge: Cambridge University Press.

Russell, Jeffrey Burton. 1991. *Inventing the Flat Earth: Columbus and Modern Historians.* New York: Praeger. A fascinating study of the way in which, during the nineteenth century, the flat-earth theory was falsely attributed to the Middle Ages and became embedded in popular and intellectual thought to the present day.

Sabra, A. I. 1987. "The Appropriation and Subsequent Naturalization of Greek Science in Medieval Islam: A Preliminary Statement." *History of Science* 25, part 3, no. 69: 223–243.

———. 1994. "Science and Philosophy in Medieval Islamic Theology." In *Zeitschrift für Geschichte der Arabisch-Islamischen Wissenschaften* 9:1–42.

Sambursky, Samuel. 1973. "John Philoponus." In *DSB.* Vol. 7, pp. 134–139.

Sarton, George. 1952. *A History of Science: Ancient Science Through the Golden Age of Greece.* Cambridge, MA: Harvard University Press.

———. 1959. *A History of Science: Hellenistic Science and Culture in the Last Three Centuries B.C.* Cambridge, MA: Harvard University Press.

Shakir, M. H. 1999. *The Qur'an.* Trans. M. H. Shakir, 11th ed. Elmhurst, NY: Tahrike Tarsile Qur'an, Inc., Publishers and Distributors of Holy Qur'an.

Sharpe, William D. 1973. "Isidore of Seville." In *DSB*. Vol. 7, pp. 27–28.

Sider, Robert D., ed. 2001. *Christian and Pagan in the Roman Empire: The Witness of Tertullian*. Washington, DC: Catholic University of America Press.

Singer, Charles. 1941. *A Short History of Science to the Nineteenth Century*. Oxford: Clarendon Press.

Sorabji, Richard. 1983. *Time, Creation and the Continuum: Theories in Antiquity and the Early Middle Ages*. Ithaca, NY: Cornell University Press. The lengthy sections on time and creation are relevant to the themes of science and religion.

Sorabji, Richard, ed. 1987. *Philoponus and the Rejection of Aristotelian Science*. Ithaca: Cornell University Press. A valuable collection of articles on John Philoponus and his negative reaction to significant aspects of Aristotle's natural philosophy.

Southern, R. W. 1953. *The Making of the Middle Ages*. New Haven, CT: Yale University Press. The focus of this study is the formative period of Western Europe from the late tenth century to the early thirteenth. Southern regards the eager acceptance of Aristotelian logic in the eleventh and twelfth centuries as of vital importance for the intellectual development of Western Europe.

Stahl, William H. 1962. *Roman Science: Origins, Development, and Influence to the Later Middle Ages*. Madison: University of Wisconsin Press. In Stahl's study, Roman science is basically the quadrivium of arithmetic, geometry, astronomy, and music. The study is largely about authors and the handbooks and encyclopedic treatises they composed during the late Roman Empire that were still influential in the Middle Ages.

———. 1971a. "Calcidius." In *DSB*. Vol. 3, pp. 14–15.

———. 1971b. *Martianus Capella and the Seven Liberal Arts*. Vol. 1: The *Quadrivium of Martianus Capella: Latin Traditions in the Mathematical Sciences 50 B.C.–A.D. 1250*. New York: Columbia University Press.

———. 1974. "Martianus Capella." In *DSB*. Vol. 9, pp. 140–141.

Stiefel, Tina. 1985. *The Intellectual Revolution in Twelfth-Century Europe*. New York: St. Martin's Press. Describes the new emphasis on rational explanations of natural phenomena.

Swain, Joseph Ward. 1950. *The Ancient World*. Vol. 2: *The World Empires: Alexander and the Romans After 334 B.C.* New York: Harper & Brothers.

Sylla, Edith Dudley. 1975. "Autonomous and Handmaiden Science: St. Thomas Aquinas and William of Ockham on the Physics of the Eucharist." In John E. Murdoch and Edith Dudley Sylla, eds., *The Cultural Context of Medieval Learning*. Dordecht, Holland: D. Reidel, pp. 349–396. Shows the major differences in the ways that Thomas Aquinas, in the thirteenth century, and William of Ockham, in the fourteenth century, applied natural philosophy, or physics, to the theological doctrine of the Eucharist.

Synan, Edward. 1980. "Introduction: Albertus Magnus and the Sciences." In James A. Weisheipl, ed., *Albertus Magnus and the Sciences: Commemorative Essays 1980*. Toronto: Pontifical Institute of Mediaeval Studies, pp. 1–12.

Taylor, A. E. 1955. *Aristotle*. New York: Dover Publications. (Repr. 1919 ed.)

Tertullian. 1885. *Tertullian's On Prescription Against Heretics*. Trans. Peter Holmes. In Alexander Roberts and James Donaldson, eds., *The Ante-Nicene Fathers*. Vol. 3. Buffalo, NY: The Christian Literature Publishing Company.

Thomas, Phillip Drennon. 1971. "Cassiodorus Senator." In *DSB*. Vol. 3, pp. 109–110.

Thomas Aquinas. 1948. *Introduction to Saint Thomas Aquinas.* Ed. Anton C. Pegis. New York: Modern Library.

———. 1967. *Summa theologiae*. 60 vols. Ed. and trans. T. Gilby et al. New York: Blackfriars/McGraw-Hill, 1964–1976: Vol. 10: Cosmogony (1a. 65–74); Trans. William A. Wallace.

Thompson, James Westfall, and Edgar Nathaniel Johnson. 1937. *An Introduction to Medieval Europe 300–1500*. New York: W.W. Norton. An older but still valuable general account of medieval history, which also includes helpful genealogical tables of royal families.

Thorndike, Lynn. 1923–1958. *A History of Magic and Experimental Science.* 8 vols. New York: Columbia University Press. A monumental study that seeks to show an intimate interrelationship between magic (including much on alchemy and astrology) and science. The first four volumes cover the Middle Ages, extending from the period of the Roman Empire to the end of the fifteenth century. (The last four volumes are devoted to the sixteenth and seventeenth centuries.) Many natural philosophers and theologians are discussed in detail, often in separate chapters or sections.

———. 1944. *University Life and Records in the Middle Ages*. New York: Columbia University Press. A comprehensive collection of 176 source documents translated from Latin to English. All aspects of university life are represented, including a number of readings relevant to the

tensions that existed between science and religion at the University
of Paris.

Wallace-Hadrill, D. S. 1968. *The Greek Patristic View of Nature*. Manchester, UK:
Manchester University Press. Describes the opinions and interpretations of some forty church fathers on a variety of problems associated
with the creation and nature. The fathers discussed issues in cosmology, astronomy, and medicine. Contains three indexes, one on all of
the fathers, another on non-Christian writers, and, finally, one on the
numerous subjects mentioned.

Waterlow, Sarah. 1982. *Nature, Change, and Agency in Aristotle's "Physics"*. Oxford: Clarendon Press.

Watt, W. Montgomery. 1953. *The Faith and Practice of al-Ghazali*. London:
George Allen and Unwin.

————. 1985. *Islamic Philosophy and Theology: An Extended Survey*. Edinburgh:
University of Edinburgh Press.

Weinberg, Julius. 1964. *A Short History of Medieval Philosophy*. Princeton, NJ:
Princeton University Press. An excellent study treating Christian, Islamic, and Jewish philosophers. Whereas Gilson (see above) was convinced that medieval philosophy reached its climax in the thirteenth
century, Weinberg argues that the nominalists of the fourteenth century outdid their thirteenth-century predecessors.

Weisheipl, James A. 1974. *Friar Thomas d'Aquino: His Life, Thought, and Work*.
Garden City, NY: Doubleday.

William of Conches. 1997. *A Dialogue on Natural Philosophy (Dragmaticon
Philosophiae)*. Trans. Italo Ronca and Matthew Curr. Notre Dame, IN:
University of Notre Dame Press.

Wolfson, Harry A. 1929. *Crescas' Critique of Aristotle: Problems of Aristotle's
"Physics" in Jewish and Arabic Philosophy*. Cambridge, MA: Harvard
University Press.

Index

About the Author

EDWARD GRANT is Distinguished Professor Emeritus of History and Philosophy of Science at Indiana University, Bloomington. He is the author or editor of ten books, including *The Foundations of Modern Science in the Middle Ages* and *God and Reason in the Middle Ages*.